The JOY of MATHEMATICS

The
JOY of
MATHEMATICS

Marvels, Novelties, and
Neglected Gems That Are Rarely Taught in
Math Class

ALFRED S. POSAMENTIER
and Robert Geretschläger,
Charles Li, and Christian Spreitzer

Prometheus Books

59 John Glenn Drive
Amherst, New York 14228

Published 2017 by Prometheus Books

Cover images © Shutterstock
Cover design by Nicole Sommer-Lecht
Cover design © Prometheus Books

Unless otherwise indicated, all interior images are by the authors or contributors.

Inquiries should be addressed to
Prometheus Books
59 John Glenn Drive
Amherst, New York 14228
VOICE: 716–691–0133
FAX: 716–691–0137
WWW.PROMETHEUSBOOKS.COM

21 20 19 18 17   5 4 3 2 1

Library of Congress Cataloging-in-Publication Data

Names: Posamentier, Alfred S. | Geretschläger, Robert | Li, Charles, 1981- | Spreitzer,
    Christian, 1979-.
Title: The joy of mathematics : marvels, novelties, and neglected gems that are rarely
    taught in math class / by Alfred S. Posamentier, Robert Geretschläger, Charles Li,
    and Christian Spreitzer.
Description: Amherst, New York : Prometheus Books, 2017. |
    Includes bibliographical references and index.
Identifiers: LCCN 2017008896 (print) | LCCN 2017019495 (ebook) |
    ISBN 9781633882980 (ebook) | ISBN 9781633882973 (pbk.)
Subjects: LCSH: Mathematics—Popular works.
Classification: LCC QA93 (ebook) | LCC QA93 .P6674 2017 (print) | DDC 510—dc23
LC record available at https://lccn.loc.gov/2017008896

Printed in the United States of America

CONTENTS

INTRODUCTION

For many decades the mathematics curriculum has been stocked with lots of essential building blocks to enable the student to navigate properly such disciplines as science, finance, engineering, architecture, and everyday life—just to name a few! With much to cover and the stress of moving ahead at a steady clip, there are many interesting and important mathematics concepts, topics, and applications that rarely ever get mentioned in the classroom.

When students are presented with finance applications in the school curriculum, such as computing the effect of interest on a given principal in the bank, there is a missed chance to enrich them with a mathematical peculiarity such as the "Rule of 72," which enables one to determine how much time is required for a bank account to double the deposited money at a specified interest rate.

There are amazingly simple geometric phenomena that are rarely ever shown in classrooms, simply for lack of time. These might include some of the special yet straightforward characteristics of quadrilaterals inscribed in a circle, such as the incredible relationship between the diagonals of a quadrilateral and its sides, where the product of the diagonals is equal to the sum of the products of the opposite sides. Remember, this is only true when the quadrilateral's four vertices lie on the same circle.

Another opportunity missed is to show how the random placement of a point inside an equilateral triangle shares a common property with any other point in that triangle, namely that the sum of the perpendiculars to each of the three sides is always the same.

There are everyday applications that regularly seem to escape classroom presentations. For example, mechanical algorithms to multiply numbers mentally. Even though there are electronic calculators readily available, the facility of manipulating numbers mentally is clearly an asset that seems to be seen as relatively unimportant in today's technological world. Here we choose to fill in this void.

A clever application of the Fibonacci numbers—perhaps the most ubiquitous numbers in our culture—allows us to convert from miles to kilometers (or the reverse) mentally. This is particularly useful for Americans traveling outside of their country's borders, when they will have to convert distance from kilometer-indicating signs to the more familiar mileage measures.

There are many uses of algebra that can explain many mathematical curiosities, which simply bewonder the uninformed. For example, most teachers will show their students how to look at a number and determine if it is divisible by 3, but they will not take the time to explain why this "trick" works. We believe that knowing *why* this works is almost as important as knowing *how* it works. The same is true for other divisibility rules that we will present in the pages that follow.

Studying conic sections is a standard part of the high school curriculum. Teachers typically show physical models but miss quite a few truly astonishing applications. Take, for example, using the rays of light emanating from a flashlight to show that they can generate conic sections. Shining a flashlight onto the ground or onto a wall at different angles allows different shapes of light to arise. The boundaries of these shapes are conic sections (assuming that the aperture of the flashlight is circular). Depending on the angle, we obtain a circle, an ellipse, a parabola, or a hyperbola. Similarly, a circular arc (being part of a building, for instance) can appear as an elliptic arc, a hyperbola, or a parabola if we look at it from different angles. Furthermore, aspects of the geometry of conic sections can be found in many architectural masterpieces. Mathematics can also explain how we create a visual depth perception and draw a picture. The concept of perspectivity, found in many famous paintings of the Italian Renaissance, has set the stage to study and perfect this concept further. Notable among the artists is Leonardo da Vinci, who also became a model for the famous German artist Albrecht Dürer. These are some of the aspects of our artistic culture that can be explained through mathematics, but, sadly, they are often neglected in the teaching of mathematics.

The topic of probability, which is gaining ever more presence in the standard curriculum today, has some truly astonishing and counterintuitive applications that all too often are not shared with students studying this topic. The famous "birthday problem" is unfortunately not presented to many classes. This "problem" offers some highly counterintuitive results. For example, it determines that the highly unanticipated probability of two people having the same birth date in a room of thirty people is 70 percent; and, perhaps more amazingly, it determines that the probability of two people having the same birth date in a group of fifty-five people is 99 percent. Such omissions clearly weaken the instructional program, so we now take the opportunity to allow the general readership to make up for previously lost chances.

The ever-present drive by teachers, who are being rated by their students' test performance, to "teach to the test" is one reason why so many mathematical wonders are omitted from the instructional program on a nationwide level. *The Joy of Mathematics* is an attempt to fill in the many gaps in the American mathematics education system and at the same time to show the average American citizen that there are lots of entertaining and useful mathematics gems that may have eluded them during their school days. We will do this in short units so as to keep the presentation crisp and intelligible. We will also use photos and diagrams extensively to enhance the attraction and function of our examples. To keep everything accessible, we have used language and content geared toward the general reader, not the math savant. As such, we have kept in mind the idea of the French mathematician Joseph Diaz

Gergonne (1771–1859) that "it is not possible to feel satisfied at having said the last word about some theory as long as it cannot be explained in a few words to any passer-by encountered in the street."[1]

We hope to provide you with a better grasp of mathematics, and above all, a greater appreciation for its usefulness, not only in its many applications, but also in exhibiting the power and beauty of the subject in its own right.

CHAPTER 1
ARITHMETIC NOVELTIES

When you think of arithmetic, you typically consider the four basic arithmetic operations. With a little more thought, you tend to tag on the square root operation as well. Unfortunately, most of our school curriculum focuses on ensuring that we have a good mechanical command of the arithmetic operations and know the number facts as best we can to service us efficiently in our everyday life. As a result, most adults are not aware of the many amazing relationships that can be exhibited arithmetically with numbers. Some of these can be extremely useful in our everyday life as well. For example, just by looking at a number and determining if it is divisible by 3, 9, or 11 can be very useful, especially if it can be done at a glance. When it involves determining divisibility by 2, we do this without much thought, by simply inspecting the last digit. We shall extend this discussion to considering divisibility by a prime number, something that clearly is not presented in the school curriculum, with which we hope to motivate the reader to investigate further primes beyond those shown here. We truly expect that the wonders that our number system holds, many of which we will present in this book, will motivate you to search for more of these curiosities along with their justifications. Some of the units in this chapter will also provide you with a deeper understanding for our number system beyond merely arithmetic manipulations. Our introduction to a variety of special numbers will generate a greater appreciation of arithmetic than the typical school courses provide. Let us begin our journey through numbers and their operations.

WHEN IS A NUMBER DIVISIBLE BY 3 OR 9?

Teachers at various grade levels often neglect to mention to students that in order to determine whether a number is divisible by 3 or 9, you just have to apply a simple rule: If the sum of the digits of a number is divisible by 3 (or 9), then the original number is divisible by 3 (or 9).

An example will best firm up your understanding of this rule. Consider the number 296,357. Let's test it for divisibility by 3 (or 9). The sum of the digits is $2 + 9 + 6 + 3 + 5 + 7 = 32$, which is not divisible by 3 or 9. Therefore, the original number, 296, 357, is not divisible by 3 or 9.

Now suppose the number we consider is 457,875. Is it divisible by 3 or 9? The sum of the digits is $4 + 5 + 7 + 8 + 7 + 5 = 36$, which is divisible by 9 (and then, of course, divisible by 3 as well), so the number 457,875 is divisible by 3 and by 9. If by some remote chance it is not immediately clear to you whether the sum of the digits is divisible by 3 or 9, then continue with this process; take the sum of the digits of your original sum and continue adding the digits until you can visually make an immediate determination of divisibility by 3 or 9.

Let's consider another example. Is the number 27,987 divisible by 3 or 9? The sum of the digits is $2 + 7 + 9 + 8 + 7 = 33$, which is divisible by 3 but not by 9; therefore, the number 27,987 is divisible by 3 and not by 9.

If this divisibility rule is mentioned in school settings, what is typically missing from the instruction of this rule is *why* it works. Here is a brief discussion about why this rule works as it does. Consider the decimal number *abcde*, whose value can be expressed in the following way:

$$N = 10^4a + 10^3b + 10^2c + 10d + e = (9 + 1)^4a + (9 + 1)^3b + (9 + 1)^2c + (9 + 1)d + e.$$

After expanding each of the binomials, we can now represent all of the multiples of 9 as $9M$ to simplify this as

$$N = [9M + (1)^4]a + [9M + (1)^3]b + [9M + (1)^2]c + [9 + (1)]d + e.$$

Then, factoring out $9M$, we get $N = 9M[a + b + c + d] + a + b + c + d + e$, which implies that the divisibility of N by 3 or 9 depends on the divisibility of $a + b + c + d + e$ by 3 or 9, which is the sum of the digits.

As you can see, things become so much better understood and appreciated when the reason for a "rule" is presented.

WHEN IS A NUMBER DIVISIBLE BY 11?

When a teacher shows the class something that is not directly specified in the school curriculum, it often generates some enjoyment and can be motivating. Take, for example, a method of determining whether a number is divisible by 11, without actually carrying out the division process. The problem is easily solved if you have a calculator at hand, but that is not always the case. Besides, there is such a clever "rule" for testing for divisibility by 11 that it is worth knowing just for its cleverness.

The rule is quite simple: If the difference of the sums of the alternate digits is divisible by 11, then the original number is also divisible by 11. That sounds a bit complicated, but it really isn't. Finding the sums of the alternate digits means that you begin at one end of the number, and you take the first, third, fifth, etc., digits and add them together. Then you add the remaining (even-placed) digits. Subtract the two sums and inspect for divisibility by 11.

This rule is probably best demonstrated through an example. Suppose we test 918,082 for divisibility by 11. We begin by finding the sums of the alternate digits: $9 + 8 + 8 = 25$ and $1 + 0 + 2 = 3$. Their difference is $25 - 3 = 22$, which is divisible by 11, and so the number 918,082 is divisible by 11. We should point out that if the difference of the sums is equal to zero, then we can conclude that the original number is divisible by 11, since zero is divisible by all numbers. We see this in the following example: testing the number 768,614 for divisibility by 11, we find that the difference of the sums of the alternate digits $(7 + 8 + 1 = 16$ and $6 + 6 + 4 = 16)$ is $16 - 16 = 0$, which is divisible by 11. Therefore, we can conclude that 768,614 is divisible by 11.

In case you may be wondering why this technique works, we offer the following. Consider the decimal number $N = abcde$, which then can be expressed as

$$N = 10^4a + 10^3b + 10^2c + 10d + e = (11 - 1)^4a + (11 - 1)^3b + (11 - 1)^2c + (11 - 1)d + e.$$

This can be written as

$$N = [11M + (-1)^4]a + [11M + (-1)^3]b + [11M + (-1)^2]c + [11 + (-1)]d + e,$$

where, after expanding each of the binomials, $11M$ represents the terms which are multiples of 11 written together. Factoring out the $11M$ terms, we get $N = 11M[a + b + c + d] + a - b + c - d + e,$ which leaves us with an expression that would be divisible by 11, but only if this last part of the previous expression is divisible by 11, namely, $a - b + c - d + e = (a + c + e) - (b + d)$, which just happens to be the difference of the sums of the alternate digits. This is a handy little "trick" that can also enhance your understanding of arithmetic. By the way, another way of looking at this trick is to say that the number 24,847,291 is divisible by 11 if and only if we obtain a number that is divisible by 11; let's see what we get: $2 - 4 + 8 - 4 + 7 - 2 + 9 - 1 = 15.$ Therefore, since the difference of the sums was 15, which is not divisible by 11, we know that 24,847,291 is not divisible by 11.

DIVISIBILITY BY PRIME NUMBERS

In today's technological world, arithmetic skills and competencies seem to be relegated to a back burner, since a calculator is so easily available. We can assume that most adults can determine when a number is divisible by 2 or by 5, simply by looking at the last digit (i.e., the units digit) of the number. That is, if the last digit is even (such as 2, 4, 6, 8, 0), then the number itself will be divisible by 2. Furthermore, if the number formed by the last two digits is divisible by 4, then the original number itself is divisible by 4. Also, if the number formed by the last three digits is divisible by 8, then the original number itself is divisible by 8. This rule can be extended to divisibility by higher powers of 2 as well.

Similarly, for the number 5: If the last digit of the number being inspected for divisibility by 5 is either a 0 or 5, then the number itself will be divisible by 5. If the number formed by the last two digits is divisible by 25, then the original number itself is divisible by 25. This is analogous to the rule for powers of 2. Have you guessed what the relationship here is between powers of 2 and 5? Yes, they are the factors of 10, the basis of our decimal number system.

Having completed in the previous discussions, the nifty techniques for determining whether a number is divisible by the primes 3, 9, and 11, the question then is: Are there also rules for divisibility by other prime numbers? Let's consider divisibility rules by prime numbers.

Aside from the potential usefulness of being able to determine whether a number is divisible by a prime number, the investigation of such rules will provide for a better appreciation of mathematics, that is, divisibility rules provide an interesting "window" into the nature of numbers and their properties. Although this is a topic that is typically neglected from the school curriculum, it can prove useful in everyday life.

The smallest prime number that we have not yet discussed in our quest for divisibility rules is the number 7. As you will soon see, some of the divisibility rules for prime numbers are almost as cumbersome as an actual division algorithm, yet they are fun, and, believe it or not, can come in handy. As we begin our quest for divisibility rules for the early prime numbers, we will begin with the following rule for divisibility by 7.

The rule for divisibility by 7: Delete the last digit from the given number, and then subtract twice this deleted digit from the remaining number. If the result is divisible by 7, then the original number is divisible by 7. This process may be repeated until we reach a number that we can visually inspect as one that is divisible by 7.

Let's consider an example to see how this rule works. Suppose we want to test the number 876,547 for divisibility by 7. Begin with 876,547 and delete its units digit, 7, and subtract its double, 14, from the remaining number: $87,654 - 14 = 87,640$. Since we cannot yet visually inspect the resulting number for divisibility by 7, we continue the process. We delete the units digit, 0, from the previously resulting number 87,640, and subtract its double (which is still 0) from the remaining number to get $8,764 - 0 = 8,764$. It is unlikely that we can visually determine whether this number, 8,764, is divisible by 7, so we continue the process. Again, we delete the last digit, 4, and subtract its double, 8, from the remaining number to get $876 - 8 = 868$. Since we still cannot visually inspect the resulting number, 868, for divisibility by 7, we again continue the process.

Continuing with the resulting number, 868, we once again delete its units digit, 8, and subtract its double, 16, from the remaining number to get $86 - 16 = 70$, which is divisible by 7. Therefore, the number 876,547 is divisible by 7.

Before continuing with our discussion of divisibility of prime numbers, you might want to practice this rule with a few randomly selected numbers, and then check your results with a calculator.

Now for the beauty of mathematics! Why does this rather strange procedure actually work? To see why things work is the wonderful aspect of mathematics—it enlightens us!

To justify the technique of determining divisibility by 7, consider the various possible terminal digits (that we are "dropping") and the corresponding subtraction that is actually being done after dropping the last digit. In the chart below you will see how in dropping the terminal digit and doubling it, we are essentially subtracting a multiple of 7. That is, we have taken "bundles of 7" away from the original number. Therefore, if the remaining number is divisible by 7, then so is the original number, because you have separated the original number into two parts, each of which is divisible by 7, and therefore, the entire number must be divisible by 7.

There is another way to argue why this method always works, and you may also want to give this some thought: Removing the final digit and then subtracting twice this digit from the remaining number is equivalent to subtracting 21 times the final digit from the number and then dividing the resulting number by 10. (The latter is certainly possible, since the number resulting from the first step must terminate in the digit 0.) Since 21 is divisible by 7, and 10 is not, the resulting number is divisible by 7 if and only if the original number was divisible by 7.

Terminal Digit	Number Subtracted from Original		Terminal Digit	Number Subtracted from Original
1	$20 + 1 = 21 = 3 \cdot 7$		5	$100 + 5 = 105 = 15 \cdot 7$
2	$40 + 2 = 42 = 6 \cdot 7$		6	$120 + 6 = 126 = 18 \cdot 7$
3	$60 + 3 = 63 = 9 \cdot 7$		7	$140 + 7 = 147 = 21 \cdot 7$
4	$80 + 4 = 84 = 12 \cdot 7$		8	$160 + 8 = 168 = 24 \cdot 7$
			9	$180 + 9 = 189 = 27 \cdot 7$

The next prime number that we have not yet considered for divisibility is the number 13.

The rule for divisibility by 13: The procedure here is similar to that used for testing divisibility by 7, except that instead of subtracting *twice* the deleted digit, we subtract *nine* times the deleted digit each time.

Perhaps it is best for us to do an example applying this rule. Let us check for divisibility by 13 for the number 5,616. We begin with our starting number, 5,616, and delete its units digit, 6, and subtract nine times 6, or 54, from the remaining number to get $561 - 54 = 507$.

Since we still cannot visually inspect the resulting number for divisibility by 13, we continue the process. With this last resulting number, 507, we delete its units digit, 7, and subtract nine times this digit, 63, from the remaining number, which gives us $50 - 63 = -13$, which *is* divisible by 13; therefore, the original number, 5,616, is divisible by 13.

In this rule for divisibility by 13, you might wonder how we determined the "multiplier" to be 9. We sought the smallest multiple of 13 that ends in a 1. That was 91, where the tens digit is 9 times the units digit. Once again consider the various possible terminal digits and the corresponding subtractions in the following table.

Terminal Digit	Number Subtracted from Original	Terminal Digit	Number Subtracted from Original
1	$90 + 1 = 91 = 7 \cdot 13$	5	$450 + 5 = 455 = 35 \cdot 13$
2	$180 + 2 = 182 = 14 \cdot 13$	6	$540 + 6 = 546 = 42 \cdot 13$
3	$270 + 3 = 273 = 21 \cdot 13$	7	$630 + 7 = 637 = 49 \cdot 13$
4	$360 + 4 = 364 = 28 \cdot 13$	8	$720 + 8 = 728 = 56 \cdot 13$
		9	$810 + 9 = 819 = 63 \cdot 13$

In each case, a multiple of 13 is being subtracted one or more times from the original number. Hence, if the remaining number is divisible by 13, then the original number is divisible by 13.

Divisibility by 17: Delete the units digit and subtract *five* times the deleted digit from the remaining number until you reach a number small enough to determine its divisibility by 17.

We justify the rule for divisibility by 17 as we did for the rules for 7 and 13. Each step of the procedure subtracts a "bundle of 17s" from the original number until we reduce the number to a manageable size and can make a visual inspection for divisibility by 17.

The patterns developed in the preceding three divisibility rules (for 7, 13, and 17) should lead you to develop similar rules for testing divisibility by larger primes. The following chart presents the "multipliers" of the deleted terminal digits for various primes.

To Test Divisibility by	7	11	13	17	19	23	29	31	37	41	43	47
Multiplier	2	1	9	5	17	16	26	3	11	4	30	14

You may want to extend this chart. It's fun, and it will increase your perception of mathematics. You may also want to extend your knowledge of divisibility rules to include composite (i.e., nonprime) numbers.

Divisibility by composite numbers: A given number is divisible by a composite number if it is divisible by each of its relatively prime factors. The chart below offers illustrations of this rule. You might want to complete the chart to include composite numbers up to the number 48.

To Be Divisible by	6	10	12	15	18	21	24	26	28
The Number Must Be Divisible by	2, and 3	2, and 5	3, and 4	3, and 5	2, and 9	3, and 7	3, and 8	2, and 13	4, and 7

You now have a rather comprehensive list of rules for testing divisibility, as well as an interesting insight into elementary number theory. An interested reader may want to test these rules (to instill even greater familiarity with numbers) and try to develop rules to test divisibility by other numbers in base ten and to generalize these rules to other bases.

SQUARING NUMBERS QUICKLY

We all learned in school how to multiply two multidigit numbers using pencil and paper. However, if we want to multiply a number by itself (that is, to square the number), there exist shortcuts to get the answer. Moreover, the multiplication of any two numbers can be written as a combination of squares of sums and differences of these numbers. Hence, knowing how to add, subtract, and square numbers is actually enough to compute the product of any two numbers.

Squaring Numbers with a Last Digit of 5

Here is a quick way to square any number with a last digit of 5: We delete the last digit and we are left with some number N. Multiplying N by $N + 1$ and appending the digits 2 and 5 at the end yields the correct result.

For example, to compute 85^2, we delete 5, multiply the remaining digit, 8, by 9, giving 72, and append 25 at the end. The result is 7,225, which is 85^2.

Why does this rule work? If we let N denote the number that remains after we have dropped the last digit, then we can write the square of the number as $(10 \cdot N + 5)^2 = 100 \cdot N^2 + 100 \cdot N + 25 = 100 \cdot N \cdot (N+1) + 25$. The product $N \cdot (N+1)$ represents the amount of hundreds in the result. But by writing its numerical value in front of the digits 2 and 5, we assign the place value of a hundred to this number and, according to our little calculation, will end up with the square of the original number.

Squaring the Numbers between 40 and 60

There is also a quick way to square the numbers between 40 and 60. Perhaps you have already figured out the rule by yourself. We merely develop a proof similar to the earlier one. So here is the trick: Any number between 40 and 60 (not including 40 and 60) can be written as $50 \pm N$, where N is a single-digit number (for example, $58 = 50 + 8$ and $43 = 50 - 7$). To do this quick calculation for 57^2, we begin by adding $25 + 7 = 32$ and tagging on $7^2 = 49$ to get 3,249. By the way, the 7 comes from $57 = 50 + 7$. Similarly, to calculate 48^2, we subtract $25 - 2 = 23$ and tag on $2^2 = 4$, which we write as 04, to get 2,304. Again, we get the 2 since $48 = 50 - 2$. The reason that this works is that the square of such a number gives $(50 \pm N)^2 = 2,500 \pm 100N + N^2 = 100(25 \pm N) + N^2$, so the leading digits of $(50 \pm N)^2$ are $25 \pm N$, followed by N^2, written as a two-digit number.

Squaring Arbitrary Numbers

The two tricks we have just discussed settle the case for when the first or the last digit is a 5 as well as for all numbers between 40 and 60. But what about all the other numbers? Although the trick presented above relied on the fact that $2 \cdot 5 = 10$, we can use the same kind of reasoning to simplify computing the squares of arbitrary numbers. Let us compute the square of a number in which the last digit is less than 5, such as, for example, 73^2. It is helpful to think of it as $(70 + 3)^2 = 4,900 + 2 \cdot 210 + 9 = 5,329$. On the other hand, if the last digit is greater than 5, such as 29^2, we resort to writing this as $29^2 = (30 - 1)^2 = 900 - 2 \cdot 30 + 1 = 841$.

Summing up, when you want to square a number, you can often simplify this problem by first decomposing this number in a clever way, or making use of the digit-5 tricks presented here. Being very good at squaring numbers can also help you perform arbitrary multiplications, and it gives you a more sophisticated view of arithmetic.

SQUARES AND SUMS

Squares are rather ubiquitous in mathematics. Yet what is not very well known is that every integer number is either a square number or the sum of two, three, or four square numbers. Although conjectured by the Greek mathematician Diophantus (201–285 CE) in his book *Arithmetica*, he was not able to provide a proof to justify his belief. This astonishing fact was first proved by the French mathematician Joseph-Louis Lagrange (1736–1813). The result is known as Lagrange's *four-square theorem*, a concept unfortunately not presented in the school years.

Let's take a look at what this theorem tells us. Consider the number 18, and we will try to represent it as a sum of four or fewer squares: $18 = 3^2 + 3^2 = 4^2 + 1^2 + 1^2 = 3^2 + 2^2 + 2^2 + 1^2$. Here we have represented 18 at the sum of two, three, and four squares.

Here are a few more examples:

$$23 = 3^2 + 2^2 + 2^2 + 1^2$$
$$43 = 5^2 + 3^2 + 3^2$$
$$97 = 8^2 + 5^2 + 2^2 + 2^2$$

An interested reader may want verify this unusual result with other numbers.

USING SQUARES TO MULTIPLY ARBITRARY NUMBERS

If you want to multiply two numbers whose sum happens to be an even number (that is, two odd numbers or two even numbers), you can use the formula $(a + b)(a - b) = a^2 - b^2$ to reduce the problem to computing two squares and taking their difference. For example, the product $47 \cdot 59$ can be written as $(53 - 6)(53 + 6) = 53^2 - 6^2 = 2,809 - 36 = 2,773$ (and, by the way, we already discussed earlier how to very quickly compute 53^2). Remember, this trick does not work when one number is odd and the other is even. However, the product of any two such numbers of different parity can also be computed as a difference of squares by employing the formula $(a \pm b)^2 = a^2 \pm 2ab + b^2$.

We calculate $(a + b)^2 - (a - b)^2 = a^2 + 2ab + b^2 - (a^2 - 2ab + b^2)$ and obtain

$$a \cdot b = \frac{(a+b)^2 - (a-b)^2}{4},$$

which is a representation of an arbitrary product $a \cdot b$ in terms of the difference of two square numbers.

As a matter of fact, knowing from memory the squares of all numbers from, say, 1 to 20, and being aware of the formulas presented above, you can easily compute the product of any two such numbers. In this sense, remembering multiplication tables for mental arithmetic is not necessary, it suffices to remember all the squares. The Babylonian clay tablets indicate that the Babylonians used tables of squares and multiplied numbers in the way we presented here, that is, by transforming products to differences of squares.[1]

AN ALTERNATIVE METHOD FOR
EXTRACTING A SQUARE ROOT

Why would anyone want to find the square root of a number today without using a calculator? Surely, no one would do such a thing. However, you might be curious to know what is actually being done in the process of finding the square root of a number. This would allow you some independence from the calculator. The procedure typically taught in schools many years ago was somewhat rote and had little meaning to the students other than obtaining an answer. We will present a method that was generally not taught in the schools but gives a good insight into the meaning of a square root. The beauty of this method is that it really allows you to understand what is going on, unlike the algorithm that was taught in schools before the advent of calculators. This method was first published in 1690 by the English mathematician Joseph Raphson (1648–1715) in his book, *Analysis alquationum universalis*; Raphson attributed it to Sir Isaac Newton (1643–1727) in his 1671 book *Method of Fluxions*, which was not officially published until 1736. Therefore, the algorithm bears both names, the *Newton-Raphson method*.

It is perhaps best to see the method as it is used in a specific example:

Suppose we wish to find $\sqrt{27}$. Obviously, the calculator could be used here. However, you might like to guess at what this value might be. Certainly, it is between $\sqrt{25}$ and $\sqrt{36}$, or between 5 and 6, but closer to 5.

Suppose we guess at 5.2. If this were the correct square root of 27, then if we were to divide 27 by 5.2, we would get 5.2. But this is not the case; since $\frac{27}{5.2} \neq 5.2$, we know that $\sqrt{27} \neq 5.2$.

In order to find a closer approximation, we will calculate $\frac{27}{5.2} = 5.192$. Since $27 \approx (5.2) \cdot (5.192)$, one of the factors (in this case, 5.2) must be bigger than $\sqrt{27}$ and the other factor (in this case, 5.192) must be less than $\sqrt{27}$. Hence, $\sqrt{27}$ is sandwiched between the two numbers 5.2 and 5.192; that is, $5.192 < \sqrt{27} < 5.2$. So it is plausible to infer that the average of these two numbers, that is, $\frac{5.2 + 5.192}{2} = 5.196$, is a better approximation for $\sqrt{27}$ than either 5.2 or 5.192.

This process continues, each time with additional decimal places, so that an allowance is made for a closer approximation. That is, $\frac{5.192 + 5.196}{2} = 5.194$, then $\frac{27}{5.194} = 5.19831$. Taking this another step to get an even closer approximation of $\sqrt{27}$, we continue this process: $\frac{27}{5.19831} = 5.193996$, then $\frac{5.19831 + 5.193996}{2} = 5.1961530$.

This continuous process provides insight into the finding of the square root of a number that is not a perfect square. As seemingly cumbersome as the method may be, it surely provides you with a genuine understanding about the value of a square root.

SENSIBLE NUMBER COMPARISONS

Comparing large numbers in today's technological world is something that should not be neglected in the school curriculum. There are numerous techniques for comparing numbers that are not simply written out in their typical decimal form but, rather, in exponential form. We will consider one here, just to give you an opportunity to see the kind of manipulations we can make to answer questions that initially seem impossible to decipher.

The question we could be faced with is which of the two values is greater, 31^{11} or 17^{14}? In order to answer this question, we will change these bases to numbers that can be reduced to a common base. It is clear that $31^{11} < 32^{11} = \left(2^5\right)^{11} = 2^{5 \cdot 11} = 2^{55}$. Whereas $17^{14} > 16^{14} = \left(2^4\right)^{14} = 2^{56}$. Now we can clearly see that because $2^{56} > 2^{55}$, we can conclude that $17^{14} > 31^{11}$. Because of the enormous magnitude of each of these two numbers, it would be very difficult to determine which is larger without converting these to common bases.

Another comparison of number magnitudes can be demonstrated by determining the following. Which of the two following expressions is larger, $\sqrt[9]{9!}$ or $\sqrt[10]{10!}$ (where the factorial expression $n! = 1 \cdot 2 \cdot 3 \cdot 4 \cdot 5 \cdot \ldots \cdot n$)? In this case, we will raise each of the two numbers to be compared to the ninetieth power, since 90 is the common multiple of 9 and 10.

$$\left(\sqrt[9]{9!}\right)^{90} = \left(9!\right)^{\frac{1}{9} \cdot 90} = \left(9!\right)^{10} = \left(9!\right)^{9} \cdot \left(9!\right)$$

$$\left(\sqrt[10]{10!}\right)^{90} = \left(10!\right)^{\frac{1}{10} \cdot 90} = \left(10!\right)^{9} = \left(9!\right)^{9} \cdot \left(10\right)^{9}$$

If we divide each of the two end results by the same number, in this case $(9!)^9$, we find that the remaining numbers are then $9!$ and 10^9. Since each of the nine factors of $9!$ is smaller than the nine factors of 10^9, $9!$ is clearly less than 10^9; therefore, we can conclude that $\left(\sqrt[9]{9!}\right)^{90} < \sqrt[10]{10!}$. Again you will notice how searching for commonality allows us to make comparisons more easily than actually computing these incredibly large numbers.

EUCLIDEAN ALGORITHM TO FIND THE GCD

What is the *greatest common divisor* (gcd) of 15 and 10? Most people would know intuitively that the answer is 5. This intuition is most likely built up through the study of the multiplication table and through practice with arithmetic. Going further, what is gcd(364, 270)? (Symbolically, this means the greatest common divisor, gcd, of 364 and 270.) At this point, intuition doesn't help as much as it did when considering the more familiar numbers 15 and 10. One option is to calculate the prime decompositions of both numbers and obtain the gcd by looking at the lowest powers of the distinct primes showing up in both the prime decompositions. Another method is to perform the Euclidean algorithm.

Consider two positive integers a and b, where $a > b$. We can always use long division to find the remainder when we divide a by b, that is, $a = qb + r$ where q is the quotient and r is the remainder. If we set $a = 364$ and $b = 270$, and calculate, then we have $364 = 1 \cdot 270 + 94$. The Euclidean algorithm revolves around the fact that gcd(a, b) = gcd(b, r). (Any divisor of both a and b is certainly a divisor of r, and any divisor of both b and r is certainly a divisor of a.) In our example, gcd(364, 270) = gcd(270, 94).

At this point, performing the long division of 270 divided by 94 would yield $270 = 2 \cdot 94 + 82$. If we think of $a = 270$ and $b = 94$, then notice the previous observation about gcds applies once more: gcd(270, 94) = gcd(94, 82).

This process is to be repeated until we get a remainder of 0.

$94 = 1 \cdot 82 + 12$, so we have gcd(94, 82) = gcd(82, 12).
$82 = 6 \cdot 12 + 10$, so gcd(82, 12) = gcd(12, 10).
$12 = 1 \cdot 10 + 2$, so gcd(12, 10) = gcd(10, 2).
$10 = 5 \cdot 2 + 0$, so gcd(10, 2) = gcd(2, 0).

But the gcd of 2 and 0 is 2 itself, since any integer is a factor of 0, that is, $0 = 0 \cdot n$ for any n. More explicitly, $2 = 1 \cdot 2$ and $0 = 0 \cdot 2$, showing that 2 is a divisor of both 2 and 0. Clearly, 2 is the largest divisor that can go into 2, hence $\gcd(2, 0) = 2$.

Using a string of equalities, we have:

$$\gcd(364, 270) = \gcd(270, 94) = \gcd(94, 82) = \gcd(82, 12) =$$
$$\gcd(12, 10) = \gcd(10, 2) = \gcd(2, 0) = 2.$$

Let's check this result using the prime decomposition method mentioned earlier. This gives us $364 = 2^2 \cdot 7 \cdot 13$ and $270 = 2 \cdot 3^3 \cdot 5$. The only common prime is 2, and the lowest power of 2 shown in the prime decompositions is 1, hence the $\gcd(364, 270) = 2^1 = 2$.

Why would we use the Euclidean algorithm if the prime decomposition method is available? It seems like the prime decomposition method can be faster if you can quickly compute the prime decompositions of the integers in question. The "quickly" part turns out to be the problem. For very large integers, the prime decomposition can be difficult or inefficient to compute. In fact, much of the security in commerce and the internet today depends on the difficulty of figuring out whether or not a large integer is prime. The Euclidean algorithm avoids this problem if you merely want to the find the gcd of the numbers in question.

The Euclidean algorithm is a very old and efficient algorithm that can compute the greatest common divisor of two integers. While intuition is sufficient in the cases involving relatively small integers, the Euclidean algorithm is able to leverage the knowledge of long division to find the greatest common divisors of as large a pair of integers as we desire to compute. The modest prerequisites combined with the usefulness of the algorithm ensure its lasting place in our arithmetic tool kit.

SUMS OF POSITIVE INTEGERS

You may have heard the often-told childhood story of the famous German mathematician Carl Friedrich Gauss (1777–1855), who performed a remarkable feat when he was just in elementary school. His math teacher had given the class the task of adding all the positive integers from 1 to 100. The teacher expected the task of evaluating $1 + 2 + 3 + \cdots + 100$ to keep the students, including the young Gauss, busy for some time. After all, Gauss was just a little boy! To his teacher's amazement, Gauss did the calculation in just a few seconds, and apparently he was the only one with the right answer.

Young Gauss explained that rather than adding the numbers sequentially, as the rest of his class was doing, he realized that the one hundred terms in this sum can be broken up into pairs: $1 + 100, 2 + 99, 3 + 98, 4 + 97$, and so on. There are 50 such pairs, with each pair having a sum of 101. Therefore, the total sum is $50 \cdot 101 = 5{,}050$.

Many will recall that Gauss's technique can be extended to find a formula for the sum $1 + 2 + \cdots + n = \dfrac{n(n+1)}{2}$, where n is an arbitrary positive integer. There are other simple ways to establish this formula, which may not have been shown in school, such as the following visual demonstration.

Consider the diagram shown in figure 1.1, with n boxes in the bottom row and right-side column.

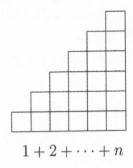

$$1 + 2 + \cdots + n$$

Figure 1.1

The "staircase" in figure 1.1 represents the sum $1 + 2 + 3 + \cdots + n$. To see this, break up the staircase into vertical columns. Looking from left to right, the left-most vertical column has 1 square, the next column over has 2 squares stacked vertically, the third column has 3 squares, and so on. The last column has n squares stacked vertically. The area of the staircase is the sum of the areas of the columns, thus the area of the staircase is $1 + 2 + 3 + \cdots + n$.

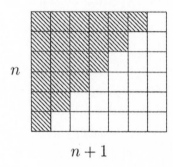

n

$n + 1$

Figure 1.2

Create a shaded copy of the inverted staircase and join it with the original one to form the rectangle shown in the figure 1.2. The rectangle has area $n(n + 1)$. The shaded and unshaded "staircases" have the same area. Thus, dividing the rectangle's area in half, each staircase has an area of $\frac{n(n+1)}{2}$. Recall that the staircase represents the sum of the integers from 1 to n. We can, therefore, conclude that $1 + 2 + \cdots + n = \frac{n(n+1)}{2}$.

This sum formula can be demonstrated in different ways, the most famous of which is probably the one that the young Gauss used. For those who prefer the visual version, the staircase method provides another elegant way of seeing how this formula holds true.

SUMS OF ODD POSITIVE INTEGERS

A few simple calculations can sometimes be enough to reveal marvelous patterns in numbers. Notice that $1 = 1^2$, which is a perfect square, that is, a number that is equal to the product of two equal integers. Notice that $1 + 3 = 4 = 2^2$, also a perfect square. Notice that $1 + 3 + 5 = 9 = 3^2$, is again a perfect square. This pattern of squares continues as expected.

We can use the following squares shown in figure 1.3 to understand this pattern:

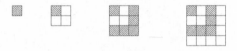

Figure 1.3

Looking from left to right, we are building progressively larger squares. At each stage, we are adding on a new L-shaped piece on the bottom right corner to get to the next larger square. The total area of the original upper left corner square and the additional L-shaped pieces equals the area of the whole square. For example, the square on the right represents $1 + 3 + 5 + 7 = 16 = 4^2$, where the 1 and 5 are areas of the shaded pieces, and the 3 and 7 are areas of the unshaded pieces. The L-shaped pieces (and the original square) all have odd areas, and the sum of these odd areas is equal to the area of the whole square, which justifies our arithmetic statement about the sum of odd integers being equal to squares.

We can also construct this pattern another way. Instead of adding odd positive integers together to get squares, let's consider the following table of squares and notice where the odd positive integers show up:

n	n^2	
0	0	0 + 1 = 1
1	1	1 + 3 = 4
2	4	4 + 5 = 9
3	9	9 + 7 = 16
4	16	16 + 9 = 25

This is a table that helps us see another pattern. To go from one row to the next, we are adding odd positive integers to the perfect squares in order to get the next perfect square. Let's focus on the third column, which represents the changes to the second column to get to the next row. Observe that from the first row to the second row, $0 + 1 = 1$. From the second row to the third row, we have $1 + 3 = 4$. From the third row to the fourth row, notice $4 + 5 = 9$. Similarly, moving along to the next row, we obtain $9 + 7 = 16$. In other words, the differences between consecutive perfect squares are the consecutive *odd* positive integers, which we might have expected—given our geometric demonstration above. Notice that by going backward from the last row up, $16 - 9 = 7$ and $9 - 4 = 5$. Going further back, we also have $4 - 1 = 3$ and $1 - 0 = 1$. Thus, starting at 0, we can add up these differences to get 16: $1 + 3 + 5 + 7 = 16$, which once again leads us to notice that the sum of consecutive odd numbers results in a square number.

Such demonstrations may be helpful if you prefer to learn visually. But do not despair, for those who prefer algebra, this idea can also be demonstrated algebraically. Consider the consecutive squares n^2 and $(n + 1)^2$, for some non-negative integer n. The difference can be computed as follows:

$$(n + 1)^2 - n^2 = n^2 + 2n + 1 - n^2 = 2n + 1.$$

Notice that the difference simplifies to $2n + 1$, which is an odd positive integer when n is non-negative.

The connection between the sum of odd positive integers and perfect squares has very humble prerequisites. The pattern itself can be seen in elementary calculations. Tables of values for the squaring function, some simple geometry, and a little bit of algebra all work together to further demonstrate this marvelous pattern in numbers.

THE REALM OF NONTERMINATING DECIMALS

Nonterminating decimals are numbers that have an infinite sequence of digits after their decimal point. They arise in various situations, some of which you are certainly familiar with. For instance, dividing one integer by another may lead to a non-terminating decimal, as well as taking the square root of some integers. The two most important mathematical constants, $e = 2.718281 \ldots$ and $\pi = 3.141592 \ldots$, are also nonterminating decimals. They have infinite sequences of digits after the decimal point, and human intuition often fails when encountering mathematical notions of infinity. There are many astonishing and remarkable facts about nonterminating decimals, some of which we believe you may not have been aware of.

Repeating Decimals

Dividing two integers using pencil and paper is a basic topic in elementary arithmetic with which we are all familiar. The result can be a terminating decimal (for example, $\frac{21}{7} = 3$, and $\frac{7}{4} = 1.75$, etc.) or a decimal with a fractional part consisting of a repeating sequence of digits (for example, $\frac{7}{3} = 2.\overline{3} = 2.333333 \ldots$). Notice that the bar over the 3 indicates that it repeats endlessly, or infinitely. The inverse problem is sometimes less emphasized, that is, given a decimal number with a fractional part, how can we construct the fraction that this number represents? This is a very easy task if it is a terminating decimal, but what if the number is something such as $1.428574285742857 \ldots$? At first sight it is not so obvious as to how to convert this number into a fraction. Unfortunately, using a pocket calculator may not be of much help either. Nevertheless, it can be done without much effort by applying a little trick. First, we need to know the "length" of the repeating part, measured in digits. For example, the repeating part of $x = 1.\overline{42857}$ is 5 digits long. Now multiply x by a corresponding power of ten and subtract x from the result. This leads directly to the desired representation of x as a fraction. In our example we would have $x = 1.\overline{42857}$, then, since the repetition is 5 digits, we multiply by 10^5 to get: $100,000x = 142857.\overline{42857}$.

Subtracting the first equation from the second yields $100,000x - x = 142,856$. So, we obtain $99,999x = 142,856$ and, thus, $x = \dfrac{142856}{99999}$, which cannot be reduced any further. Applying this conversion procedure to the repeating decimal $0.\overline{9} = 0.999999\ldots$ reveals that the decimal representation of a number is not always unique: For $x = 0.\overline{9}$, we would get $10x = 9.\overline{9}$, and by subtracting we get $9x = 9$, or $x = 1$. Thus implying that $0.\overline{9} = 1$. This means that $0.\overline{9}$ is really just another representation of 1. Intuitively you might be tempted to believe that $0.\overline{9}$ should be a tiny little bit smaller than 1, but it isn't (as we just proved). We cannot always trust our intuition when dealing with infinite sequences.

Irrational Numbers

If a nonterminating decimal does not have a repeating pattern of numbers, then it cannot be written as a fraction. Thus, it is not a rational number and therefore, such numbers are then called irrational. Euler's number e and π are both examples of irrational numbers. The square root of a natural number is irrational whenever the number is not a square number; for example, $\sqrt{2}, \sqrt{3}, \sqrt{5}, \sqrt{6}, \sqrt{7}, \sqrt{8}, \sqrt{10}$, and $\sqrt{11}$ are all irrational numbers. They have infinite nonrepeating sequences of digits after the decimal point. Clearly, it is not possible to know all decimal values of π, e, or $\sqrt{2}$, since they go on infinitely far and without any obvious pattern.

Mathematicians are always in search of patterns among the digits of the decimal approximation of π, where sometimes patterns can be established. For example, the British mathematician John Conway (1937–) has indicated that if you separate the decimal value of π into groups of ten places, the probability of each of the ten digits appearing in any of these blocks is about 1 in 40,000. Yet he noted that it does occur in the seventh such group of ten places, as you can see from the grouping below:

π = 3.1415926535 8979323846 2643383279 5028841971
6939937510 5820974944 **5923078164** 0628620899 8628034825
3421170679 8214808651 3282306647 0938446095 5058223172
5359408128 . . . .

However, accompanied by the advances in computational power, the record precision of numerical approximations to π is steadily increasing. Recently, the first 22,459,157,718,361 (about 22.5 trillion) decimal places of π have been computed, but most likely at your reading, this record will already have been broken. Irrational numbers are a challenge not only for computers but also for people who like to memorize numbers. Believe it or not, there are even separate world-ranking lists for memorizing digits of π, e, and $\sqrt{2}$. For example, the current record (as of March 2017) for π is held by Suresh Kumar, who was able to recite its first 70,030 digits, and likely this will have been superseded at this reading.[2] On the other hand, it is often overlooked that not all irrational numbers have decimal representations that are hard to memorize. For example, consider the number 0.123456789101 112131415161718192021222324252627282930313233334 . . . .

Can you see how this sequence continues? This number is called Champernowne's constant, named after the English mathematician D. G. Champernowne (1912–2000), who published it as an undergraduate student in 1933. Its fractional part is obtained by concatenating all positive integers in order—a sequence everyone can write down immediately. Observe that any finite sequence of numbers will occur somewhere in the infinite sequence of digits of this number. In fact, any finite sequence of digits will even occur infinitely often in Champernowne's number. For example, if we would use a sequence of numbers to represent the code imprinted in the sequence of nucleobases in a DNA molecule, then the exact sequence of numbers representing your own DNA molecule would occur somewhere in the endless digits of Champernowne's number. Of course, the same would be true for the DNA molecule of any other organism that lives, or has ever lived, on Earth. This may be hard to believe, but it is just a simple consequence of the concept of infinity and the definition of Champernowne's constant (and it does not imply that this number has any special meaning).

These strange features of the seemingly all-encompassing Champernowne's constant once again illustrate that infinite sequences (and mathematical notions of infinity in general) are completely different from everything we experience in real life. If you are not used to dealing with such concepts, then you may be puzzled by some of the counterintuitive facts accompanying them.

ATOMS IN THE UNIVERSE OF NUMBERS

In chemistry class, we are introduced to the periodic table of elements, which contains all known extant chemical elements. Some of them have only been produced in laboratories and do not occur in nature. As far as we know, all visible matter in the universe is made up of 94 different natural chemical elements, from the lightest element, hydrogen, to the heaviest, plutonium. Although 24 even heavier elements have been produced artificially, they have extremely short half-lives and could not be observed in nature. The 94 natural chemical elements, representing 94 different sorts of atoms, can be regarded as the elementary building blocks of our world. Every piece of matter can be decomposed into a finite number of atoms, each belonging to one of the different elements. For instance, a droplet of water is made up of a vast number of water molecules, each of which is formed by two hydrogen atoms and one oxygen atom. Thus, the droplet contains some number of oxygen atoms and twice that number of hydrogen atoms. Similarly, we can decompose every isolated conglomerate of matter into individual atoms and sort these atoms according to the chemical elements they represent. The word *atom* was created by ancient Greek philosophers and meant the "indivisible," that is, the smallest unit of matter. In ancient Greece, philosophy, physics, and mathematics were not separate disciplines; they all belonged together as "natural philosophy," meaning the philosophical study of nature and the physical universe. Ancient Greek philosophers also noticed that smallest, indivisible units exist in the world of numbers as well. They are now called prime numbers—from Latin *numerus primus* (meaning "first numbers"). A prime number is a natural number that has exactly two natural numbers as divisors (the number itself and 1). Bear in mind that 1 is not a prime number by this definition, since it has no other divisor than itself.

Just as a piece of matter can be decomposed into individual atoms, each representing a certain chemical element, every integer greater than 1 can be decomposed into indivisible factors, each of them representing a certain prime number. But while there are only 94 different natural chemical elements in nature, there are infinitely many prime numbers. Notwithstanding the infinite number of primes, the decomposition of an integer into prime factors is unique, just like the decomposition of matter into atoms. This important statement is called the *fundamental theorem of arithmetic*. Euclid of Alexandria (fl. 300 BCE) gave a proof of this theorem in his famous book Στοιχεῖα (Greek: *Stoicheia*), which is now known as Euclid's *Elements*. Although the proof is elementary from a mathematical viewpoint, we will not present it here, since it requires some special notation with which not all readers would be familiar. We will instead try to motivate and explain the result on a heuristic level.

The Fundamental Theorem of Arithmetic

Suppose we are given an arbitrary integer greater than 1. Then either this number is a prime number, meaning it has no divisors other than 1 and itself, or it is not a prime number. If it is prime, the number itself represents its unique prime factor decomposition. However, if the number is not prime, then we can always break it up into prime factors, thereby obtaining a product of prime numbers representing the given number. Such a nonprime number is called a *composite number*. The following table shows the prime factor decompositions of the forty smallest integers greater than 1, which are composite numbers. The prime numbers, which are "missing" from the list are the numbers 2, 3, 5, 7, 11, 13, 17, 19, 23, 29, 31, 37, 41, 43, 47, and 53, and they are identical to their prime factorizations.

$4 = 2^2$	$20 = 2^2 \cdot 5$	$33 = 3 \cdot 11$	$46 = 2 \cdot 23$
$6 = 2 \cdot 3$	$21 = 3 \cdot 7$	$34 = 2 \cdot 17$	$48 = 2^4 \cdot 3$
$8 = 2^3$	$22 = 2 \cdot 11$	$35 = 5 \cdot 7$	$49 = 7^2$
$9 = 3^2$	$24 = 2^3 \cdot 3$	$36 = 2^2 \cdot 3^2$	$50 = 2 \cdot 5^2$
$10 = 2 \cdot 5$	$25 = 5^2$	$38 = 2 \cdot 19$	$51 = 3 \cdot 17$
$12 = 2^2 \cdot 3$	$26 = 2 \cdot 13$	$39 = 3 \cdot 13$	$52 = 2^2 \cdot 13$
$14 = 2 \cdot 7$	$27 = 3^3$	$40 = 2^3 \cdot 5$	$54 = 2 \cdot 3^3$
$15 = 3 \cdot 5$	$28 = 2^2 \cdot 7$	$42 = 2 \cdot 3 \cdot 7$	$55 = 5 \cdot 11$
$16 = 2^4$	$30 = 2 \cdot 3 \cdot 5$	$44 = 2^2 \cdot 11$	$56 = 2^3 \cdot 7$
$18 = 2 \cdot 3^2$	$32 = 2^5$	$45 = 3^2 \cdot 5$	$57 = 3 \cdot 19$

The fact that every integer greater than 1 can be broken up into the product of prime numbers is not so surprising, given the definition of a prime number. A number that is not prime must have integer divisors other than 1 and itself and can therefore be factored, that is, split up into factors (e.g., $12 = 4 \cdot 3$). If any of these factors is not prime, it can also be split up into smaller factors, and so forth. Evidently, the procedure of factoring will stop when all obtained factors cannot be divided any further, that is, when they are all prime numbers (e.g., $12 = 2 \cdot 2 \cdot 3 = 2^2 \cdot 3$). So it is actually quite obvious that integers can be represented as products of prime numbers. However, the fundamental theorem of arithmetic also states that this decomposition is unique (the order of the factors is not important; e.g., $12 = 2 \cdot 2 \cdot 3 = 2 \cdot 3 \cdot 2 = 3 \cdot 2 \cdot 2$). For example, $2016 = 2^5 \cdot 3^2 \cdot 7$, and there is no other way to represent 2016 as the product of prime numbers. Thus, independent of the way we do the factorization (by using a certain algorithm or simply by using a trial-and-error strategy), we will end up with five 2s, two 3s, and one 7.

Is Prime Factorization Really Special?

Another way of looking at the fundamental theorem of arithmetic is to view it as a statement on the "composition" of integers rather than on the "decomposition": All integers greater than 1 can be constructed or "composed" by multiplying prime numbers, and for each integer there is only one special composition of primes representing this integer. Hence, prime numbers can truly be regarded as the basic building blocks (or the "atoms") of integers.

One could argue that every integer can as well be constructed by adding a unique number of 1s; for example, $12 = 1 + 1 + 1 + 1 + 1 + 1 + 1 + 1 + 1 + 1 + 1 + 1$, so 1 could then be called the building block of all integers. Yet there is a decisive difference to prime factorization: If we want a sum of 1s to become 12, we also need twelve 1s. More generally, if we want to represent an integer N as a sum of 1s, we need N of them. So we actually need N to "construct" N as a sum of 1s. As a matter of fact, representing N as a sum of 1s does not provide any additional information about N; it is essentially just another way of writing this number (such as writing a number using Roman numerals). In contrast, when multiplying primes, the prime numbers themselves "construct" the number N. For instance, to obtain 12, the prime factorization $2^2 \cdot 3$ already contains all the information—nothing more is needed.

The Atoms of Numbers

As we pointed out earlier, all molecules (and, more generally, all pieces of matter) consist of specific numbers of atoms from different chemical elements. Analogously, every integer greater than 1 consists of specific numbers of different primes. We can represent molecules by chemical formulas. For example, H_2O for the water molecule (H_2O stands for 2 hydrogen atoms and 1 oxygen atom). Furthermore, every integer greater than 1 can be represented by a unique product of primes, the prime factorization of this number. For instance, the expression $2^5 \cdot 3^2 \cdot 7$ plays the same role for the number 2016 as the chemical formula H_2O does for the water molecule.

Applications of Prime Factorization

For more than 2,000 years, prime numbers and the fundamental theorem of arithmetic seemed to be of little practical value. This changed with the advent of computer technology. The fundamental theorem of arithmetic gives no information about how to obtain an integer's prime factorization; it only guarantees its existence. While there exist systematic methods to decompose an integer into prime factors, the number of operations required in such procedures increases very rapidly with the number of digits of the given integer. Factoring many-digit integers is only possible with the help of computers. However, if the number to be factored is sufficiently large (say, several hundred digits), prime factorization is virtually impossible, even for the most powerful super-computers. It would simply take too much time. Many public-key cryptosystems for secure data transmission are based on this fact. In public-key cryptography, each user has a pair of cryptographic keys—a public encryption key and a private decryption key. The public encryption key may be widely distributed, while the private decryption key is known only to its owner. A typical application of public-key cryptography are digital signatures used in financial transactions to demonstrate the authenticity of a digital message. There is a mathematical relation between the encryption key and the decryption key, but calculating the private key from the public key is unfeasible, since it would involve finding the prime factors of a very large number. Thus, the safety of such cryptosystems directly depends on the mathematical difficulty of factoring large numbers. Interestingly, a hypothetical quantum computer (that is, a computation system making direct use of quantum-mechanical phenomena) could factor even large integers quickly. This was shown by the American mathematician Peter Williston Shor (1959–), who developed an algorithm for quantum computers that runs exponentially faster than the best currently known algorithm running on a classical computer. However, it has not yet been proved that there does not exist an efficient prime factorization algorithm for classical computers. It's just that nobody has yet found one.

FUN WITH NUMBER RELATIONSHIPS

Today's the school curriculum seems very heavily focused on testing students. This has many teachers gearing their instruction toward passing these examinations. It would be refreshing to encourage teachers to entertain students with number peculiarities. Frankly, the one advantage of taking time to show these number relationships is to demonstrate the beauty that lies well hidden in our number system, which should motivate students toward embracing mathematics. These unexpected relationships are boundless in their manifestations. We will present some of these here as a form of entertainment with the hope that you will then be motivated to seek other such clever patterns.

Let's begin by considering numbers where we will raise each of the digits of the number to the third power and show that their sum is equal to the original number:

$407 = 4^3 + 0^3 + 7^3$
$153 = 1^3 + 5^3 + 3^3$
$371 = 3^3 + 7^3 + 1^3$

A similar situation can be shown for fourth and fifth powers, as in the following examples:

$1,634 = 1^4 + 6^4 + 3^4 + 4^4$
$4,150 = 4^5 + 1^5 + 5^5 + 0^5.$

There are many other numbers that can be expressed as the sum of its digits each taken to the same power. We invite you to begin your search. But first we will start you off with a clue to one such number: 8,208, which can be expressed as the sum of its digits taken to a power. We leave it to you to determine to which power these digits need to be raised.

We can do this again, but this time we have two numbers related to each other in a similar fashion as above, that is, each number can be expressed as the sum of the digits of the other number, each taken to the same power. In our example that follows, the number 136 and the number 244 have this relationship:

$136 = 2^3 + 4^3 + 4^3$; now we take these bases to form the number $244 = 1^3 + 3^3 + 6^3$, whose bases determine the original number.

Another unusual arrangement of powers can be seen from the value of 204^2, which can be shown to be equal to three consecutive numbers taken to third power: $204^2 = 23^3 + 24^3 + 25^3$.

Taking this a step further, we consider the number $8,000 = 20^3$, which can also be expressed as a sum of consecutive cubes—this time four cubes—as follows: $20^3 = 11^3 + 12^3 + 13^3 + 14^3$.

There are other numbers that can also be expressed as the sum of consecutive numbers taken to the same power. Before you search for others, we will further entice you with one more example:

$4,900 = 70^2 = 1^2 + 2^2 + 3^2 + 4^2 + 5^2 + 6^2 + \cdots + 20^2 + 21^2 + 22^2 + 23^2 + 24^2$.

Now to consider consecutive exponents. There are also numbers that are equal to the sum of the digits raised to consecutive powers, such as the following:

$135 = 1^1 + 3^2 + 5^3$
$175 = 1^1 + 7^2 + 5^3$
$518 = 5^1 + 1^2 + 8^3$
$598 = 5^1 + 9^2 + 8^3$

Expressing numbers as the sum of powers also provides us some further entertainment. Some of these are quite ingenious. For example, in 1772 the famous Swiss mathematician Leonhard Euler (1707–1783) discovered that $159^4 + 158^4 = 635,318,657 = 133^4 + 134^4$. This can be extended to considering a number, 6,578, which can be expressed as a sum of three fourth-powers in two distinct ways: $6,578 = 1^4 + 2^4 + 9^4 = 3^4 + 7^4 + 8^4$. This happens to be the smallest number for which this is possible.

We can also express certain numbers as a sum of equal powers—less than the fourth power:

$65 = 8^2 + 1^2 = 7^2 + 4^2$
$125 = 10^2 + 5^2 = 11^2 + 2^2 = 5^3$
$250 = 5^3 + 5^3 = 13^2 + 9^2 = 15^2 + 5^2$
$251 = 1^3 + 5^3 + 5^3 = 2^3 + 3^3 + 6^3$

Another unusual sum of powers, where each of the numbers is taken to the same power as the original number, is shown here: $102^7 = 12^7 + 35^7 + 53^7 + 58^7 + 64^7 + 83^7 + 85^7 + 90^7$.

Perhaps you will be amused to find a number that is equal to the sum of all of the two-digit numbers that can be formed by the digits of the original number. Our example will be with the number 132, which also happens be the smallest number for which this is possible, namely, $132 = 12 + 13 + 21 + 23 + 31 + 32$.

There are boundless curious patterns that exist in our number system. Unfortunately, there rarely seems to be a really good opportunity for them to be presented to students or the general public. However, the fun of discovering these unusual relationships adds to an aspect of mathematics that can be entertaining and also enlightening as you search for more such patterns.

FRIENDLY NUMBERS

As we have indicated, unfortunately, there is hardly enough time in the course of learning mathematics in school to show some of the unusual properties that numbers have, and which have given mathematicians throughout the centuries material to further investigate. We are all aware that certain numbers have properties in common. For example, even numbers are all divisible by 2. We know that odd numbers are not divisible by 2. These are common relationships among numbers. There are, however, relationships between numbers that are quite unusual. One such relationship has been termed numbers that are "friendly" to each other. What could possibly make two numbers friendly? Mathematicians have decided that two numbers are to be considered friendly—or as is sometimes used in the more sophisticated literature, "amicable"—if the sum of the proper divisors[3] (or factors) of one number equals the second number *and* the sum of the proper divisors of the second number equals the first number as well. Sounds complicated? It really isn't. Just take a look at the smallest pair of friendly numbers: 220 and 284. The proper divisors (or factors) of **220** are 1, 2, 4, 5, 10, 11, 20, 22, 44, 55, and 110. Their sum is $1 + 2 + 4 + 5 + 10 + 11 + 20 + 22 + 44 + 55 + 110 = \textbf{284}$. The proper divisors of **284** are 1, 2, 4, 71, and 142, and their sum is $1 + 2 + 4 + 71 + 142 = \textbf{220}$. This shows that the two numbers can be considered a pair of *friendly numbers*.

The second pair of friendly numbers, which were discovered by the famous French mathematician Pierre de Fermat (1601–1665) is 17,296 and 18,416. In order for us to establish their friendliness relationship, we need to find all of the prime factors of each of the numbers: $17,296 = 2^4 \cdot 23 \cdot 47$, and $18,416 = 2^4 \cdot 1,151$. Then we need to create all the numbers from these prime factors as follows. The sum of the factors of 17,296 is

$$1 + 2 + 4 + 8 + 16 + 23 + 46 + 47 + 92 + 94 + 184 + 188 + 368 + 376 + 752 + 1081 + 2162 + 4324 + 8648 = 18,416.$$

The sum of the factors of 18,416 is

$$1 + 2 + 4 + 8 + 16 + 1151 + 2302 + 4604 + 9208 = 17,296.$$

Once again, we notice that the sum of the factors of 17,296, is equal to 18,416, and, conversely, the sum of the factors of 18,416, is equal to 17,296. This qualifies these two numbers to be considered a pair of friendly numbers.

There are many more such pairs of friendly numbers. The following are a few of such pairs of friendly numbers for your consideration:

1,184 and 1,210
2,620 and 2,924
5,020 and 5,564
6,232 and 6,368
10,744 and 10,856
9,363,584 and 9,437,056
111,448,537,712 and 118,853,793,424

If you are feeling ambitious, you may want to verify the above pairs' friendliness!

For the experts, the following is one method for finding pairs of friendly numbers:

Let $a = 3 \cdot 2^n$, $b = 3 \cdot 2^{n-1} - 1$, and $c = 3^2 \cdot 2^{n-1} - 1$, where n is an integer greater than or equal to 2, and a, b, and c are all prime numbers. It then follows that $2^n ab$ and $2^n c$ are friendly numbers.

We can always look for fascinating relationships between numbers. We now know what is meant by pairs of friendly numbers. With some creativity, we can establish another form of "friendliness" between numbers. Some of them can be truly mind-boggling! Take for example the pair of numbers 6,205 and 3,869.

At first glance, there seems to be no apparent relationship between these two numbers. But with some luck and imagination, we can get some fantastic results:

$6{,}205 = 38^2 + 69^2$, and $3{,}869 = 62^2 + 05^2$.

We can even find another pair of numbers with a similar relationship. Consider these:

$5{,}965 = 77^2 + 06^2$, and $7{,}706 = 59^2 + 65^2$.

Beyond the enjoyment of seeing this wonderful pattern, there isn't much mathematics in these examples. However, the relationship is truly amazing and worth noting. Again, mathematics has its hidden treasures, many of which have passed by the average math student without proper fanfare!

PALINDROMIC NUMBERS

The average school curriculum is rather limited with regard to the types of numbers that are presented to students throughout their mathematics

THE JOY OF MATHEMATICS

48

instruction. Surely, students know about odd numbers, even numbers, prime numbers, and even perfect numbers, which we will discuss later in this chapter. However, there are other kinds of numbers that have an unusual property and are often neglected, such as numbers that read the same in both directions. These numbers are called *palindromic numbers*; they read the same left to right as they do from right to left. First, recall that a palindrome can also be a word, phrase, or sentence that reads the same in both directions. Figure 1.4 shows a few amusing palindromes.

A
EVE
RADAR
REVIVER
ROTATOR
LEPERS REPEL
MADAM I'M ADAM
STEP NOT ON PETS
DO GEESE SEE GOD
PULL UP IF I PULL UP
NO LEMONS, NO MELON
DENNIS AND EDNA SINNED
ABLE WAS I ERE I SAW ELBA
A MAN, A PLAN, A CANAL, PANAMA
A SANTA LIVED AS A DEVIL AT NASA
SUMS ARE NOT SET AS A TEST ON ERASMUS
ON A CLOVER, IF ALIVE, ERUPTS A VAST, PURE EVIL; A FIRE VOLCANO

Figure 1.4

A palindrome in mathematics would be a number, such as 666 or 123321, that reads the same in either direction. For example, the first four powers of 11 are palindromic numbers:

$11^0 = 1$
$11^1 = 11$
$11^2 = 121$
$11^3 = 1331$
$11^4 = 14641$

It is interesting to see how a palindromic number can be generated from randomly selected numbers. All you need to do is to continually add a number to its reversal (that is, the number written in the reverse order of digits) until you arrive at a palindrome. For example, a palindrome can be reached with a single addition when the starting number is 23: 23 + 32 = 55, which is a palindrome. Or it might take two steps, such as when the starting number is 75: 75 + 57 = 132 and 132 + 231 = 363, which has led us to a palindrome. Or it might take three steps, such as with the starting number 86: 86 + 68 = 154, 154 + 451 = 605, and 605 + 506 = 1111, whereby we finally reach a palindromic number. If we start with the number 97, it will require six steps to reach a palindrome; but if we start with the number 98, it will require 24 steps to reach a palindrome.

Be cautioned about using the starting number 196; this one has not yet been shown to produce a palindromic number—even with over three million reversal additions. We still do not know if this starting number will ever reach a palindrome. If you were to try to apply this procedure with 196, you would eventually—at the sixteenth addition—reach the number 227,574,622, which you would also reach at the fifteenth step of the attempt to get a palindrome when the starting number is 788. This would then imply that applying the procedure to the number 788 has also never been shown to reach a palindrome. As a matter of fact, among the first 100,000 natural numbers, there are 5,996 numbers for which we have not yet been able to show that the procedure of reversal additions will lead to a palindrome. Some of these are 196, 691, 788, 887, 1675, 5761, 6347, and 7436.

Using this procedure of reverse-and-add, we find that some numbers yield the same palindrome in the same number of steps, such as with the numbers 554, 752, and 653, which all produce the palindrome 11011 in three steps. In general, all integers, where the corresponding digit pairs symmetric to the middle 5 have the same sum, will produce the same palindrome in the same number of steps. However, there are other integers that produce the same palindrome, yet in a different number of steps, such as the number 198, which with repeated reversals and additions will reach the palindrome 79497 in five steps, while the number 7299 will reach this number in two steps.

For a two-digit number ab with digits $a \neq b$, the sum of its digits, $a + b$, determines the number of steps needed to produce a palindrome. Clearly, if the sum of the digits is less than 10, then only one step will be required to reach a palindrome—for example, $25 + 52 = 77$. If the sum of the digits is 10, such as for the starting number 73, then we get $73 + 37 = 110$. In that case, $ab + ba = 110$, and $110 + 011 = 121$, and two steps will be required to reach the palindrome. The number of steps required for each of the two-digit sums 11, 12, 13, 14, 15, 16, and 17 to reach a palindromic number are 1, 2, 2, 3, 4, 6, and 24, respectively.

Had this topic been brought into the classroom as part of the curriculum, one would be able to appreciate some lovely patterns with palindromic numbers. For example, some palindromic numbers when squared also yield a palindrome. For example, $22^2 = 484$ and $212^2 = 44944$. These are special cases and do not lend themselves to any general rule. Such as, for example, if we square the palindromic number $545^2 = 297,025$, we, clearly, do not end up with a palindrome. On the other hand, as you might expect, there are also nonpalindromic numbers that, when squared, yield a palindromic number, such as $26^2 = 676$ and $836^2 = 698,896$. These are just some of the amusements that numbers provide, which are often neglected in school curricula. By including topics that are perceived to be entertaining and ultimately fun for students to experience, teachers will have taken a giant step toward motivating students not only to appreciate mathematics but also to provide a favorable dimension to a subject that is often seen as dreary, mechanical, and sometimes repetitive. You may want to search for other curiosities to further enhance your own appreciation.

We can actually take the concept of a palindromic number one step further to another kind of palindrome, one which consists of numbers composed entirely of 1s. These are called *repunits*. All of the repunit numbers with fewer than ten 1s, when squared, yield palindromic numbers, such as in the case of $1111^2 = 1234321$. There are also some palindromic numbers that, when cubed, yield again palindromic numbers. To this set belong all numbers of the form $n = 10^k + 1$, for $k = 1, 2, 3 \ldots$. When n is cubed, it yields a palindromic number that has $k - 1$ zeros between each consecutive pair of 1,3,3,1, as shown with the following examples:

$k = 1, n = 11$, we get: $11^3 = 1331$
$k = 2, n = 101$ we get: $101^3 = 1030301$
$k = 3, n = 1001$ we get: $1001^3 = 1003003001$

We can continue to generalize and get some interesting patterns. For example, when a number n consisting of three 1s and any even number of 0s symmetrically placed between the end 1s is cubed, the resulting number will be a palindrome. Some of these are the following:

$111^3 = 1367631,$
$10101^3 = 1030607060301,$
$1001001^3 = 1003006007006003001$, and
$1000100001^3 = 1000300060007000600030001.$

Taking this even a step further, we find that when n consists of four 1s and 0s in a palindromic arrangement, where the places between the 1s do not have same number of 0s, then n^3 will also be a palindrome, as we can see with the following examples:

$11011^3 = 1334996994331$ and
$10100101^3 = 1030331909339091330301$

However, when the same number of 0s appears between the 1s, then the cube of the number will not result in a palindrome, as in the following example: $1010101^3 = 1030610121210060301$. As a matter of fact, the number 2201 is the only nonpalindromic number that is less than 280,000,000,000,000 that, when cubed, yields a palindrome: $2201^3 = 10662526601$.

However, just for amusement, consider the following pattern with palindromic numbers:

$$12321 = \frac{333 \cdot 333}{1 + 2 + 3 + 2 + 1},$$

$$1234321 = \frac{4444 \cdot 4444}{1 + 2 + 3 + 4 + 3 + 2 + 1},$$

$$123454321 = \frac{55555 \cdot 55555}{1 + 2 + 3 + 4 + 5 + 4 + 3 + 2 + 1},$$

$$12345654321 = \frac{666666 \cdot 666666}{1 + 2 + 3 + 4 + 5 + 6 + 5 + 4 + 3 + 2 + 1},$$

and so on.

Naturally, there is much more to admire about palindromes, and yet even as much as we provided here would probably take a good portion of time from the school curriculum. Such an expenditure of time is merely an investment; benefits include more motivated students eager to learn more mathematics.

PRIME NUMBERS

Prime numbers are mentioned at various times throughout the school curriculum. As we explained earlier, prime numbers are clearly defined as those numbers that are divisible only by 1 and themselves. Some of the first primes are 2, 3, 5, 7, 11, 13, 17, 19, ... The question then is, is the number 1 a prime or not? It seems to meet the criterion that it is divisible by 1 and itself. However, mathematicians have chosen to eliminate

1 from the list of primes. One reason for that is that we agree that every composite number (that is, a nonprime number) can be expressed in a unique way as the product of prime numbers. For example, the number 30 can be expressed uniquely as the product of primes as $2 \cdot 3 \cdot 5$. If we allowed the number 1 to be a prime, then we would not be able to express the number 30 uniquely as the product of primes, since with the 1 included, we could represent the number 30 in a number of ways, such as: $1 \cdot 1 \cdot 1 \cdot 2 \cdot 3 \cdot 5$ or $1 \cdot 1 \cdot 2 \cdot 3 \cdot 5$. Consequently, the number 1 is not a prime. Among the list of primes there is also a unique prime, namely, the only *even* prime number, which is the number 2.

As we look at primes, we will notice that there are some primes that are reversible primes, that is, a prime number, which, when the digits are reversed, also yields a prime number. Some examples of this are 13 and 31, 17 and 71, 37 and 73, 79 and 97, 107 and 701, 113 and 311, 149 and 941, 157 and 751.

Among the palindromic numbers are some prime numbers, as you can see from the following samples: 2, 3, 5, 7, 11, 101, 131, 151, 181, 191, 313, 353, 373, 383, 727, 757, 787, 797, 919, 10301, 10501, 10601, 11311, 11411, 12421, 12721, 12821, and 13331.

Moreover, there are repunit numbers (that is, recall, numbers consisting of only 1s) that are also prime, such as 11, and 1111111111111111111, as well as 11111111111111111111111, with the next two such prime repunit numbers having large numbers of 1s, specifically, 317 digits and 1,031 digits—all 1s.

There are prime numbers that have the characteristic that any other arrangement of their digits will also produce a prime number. The first few of these are: 2, 3, 5, 7, 11, 13, 17, 31, 37, 71, 73, 79, 97, 113, 131, 199, 311, 337, 373, 733, 919, and 991. It is believed that larger such primes are repunit primes.

There are also prime numbers that remain prime numbers even when their digits are moved in a circular fashion. For example, the prime number 1,193 can have its digits "rotated" to form the following numbers: 1,931; 9,311; 3,119. Since all of these rotated-digit variations yield a prime number, we call the number 1,193 a circular prime number. Other such circular prime numbers are 2; 3; 5; 7; 11; 13; 17; 31; 37; 71; 73; 79; 97; 113; 131; 197; 199; 311; 337; 373; 719; 733; 919; 971; 991; 1,193; 1,931; 3,119; 3,779; 7,793; 7,937; 9,311; 9,377; 11,939; 19,391; 19,937; 37,199; 39,119; 71,993; 91,193; 93,719; 93,911; and 99,371.

The relationships between prime numbers has also fascinated mathematicians for centuries. One such relationship is the number of numbers between any two primes. For example, when two prime numbers are separated by only one other number, they are referred to as *twin primes*. The first such twin primes are 3 and 5, 5 and 7, 11 and 13, 17 and 19, 29 and 31, and so on. We should note that there are only two *consecutive* numbers that are both prime numbers, namely, 2 and 3, since 2 is the only even prime.

We can also find primes that allow us to "play" with them a bit. For example, there are additive primes, which are prime numbers, for which the sum of their digits is also a prime number. Some of these are 2, 3, 5, 7, 11, 23, 29, 41, 43, 47, 61, 67, 83, 89, 101, 113, and 131.

There are also prime numbers that are the sum of two consecutive squares. The first few of these so-called sum-of-consecutive-squares primes are: $1 + 4 = 5$, $4 + 9 = 13$, $16 + 25 = 41$, and the rest of the first few are: 61; 113; 181; 313; 421; 613; 761; 1,013; 1,201; 1,301; 1,741; 1,861; 2,113; 2,381; 2,521; 3,121; 3,613; 4,513; 5,101; 7,321; 8,581; 9,661; 9,941; 10,531; 12,641; 13,613; 14,281; 14,621; 15,313; 16,381; 19,013; 19,801; 20,201; 21,013; 21,841; 23,981; 24,421; and 26,681. An ambitious reader may try to discover the consecutive-square sum that yields each of these listed primes.

The school curriculum tends to define prime numbers and does very little beyond that. It turns out that this topic has almost boundless applications for entertainment and further study. Some of them are a little bit more like a game, but nonetheless harbor important mathematics. For example, there are some prime numbers that, when *any* one of their digits is changed to another value, will always result in a composite (nonprime) number. Some of these are 294,001; 505,447; 584,141; 604,171; 971,767; 1,062,599; 1,282,529; 1,524,181; 2,017,963; 2,474,431; 2,690,201; 3,085,553; 3,326,489; and 4,393,139.

How inspiring it would have been to have encountered these aspects of prime numbers during our school days!

INFINITE PRIMES

It should be well-known that there are an infinite number of prime numbers. This is something that has been mentioned any number of times in the school curriculum. However, rarely is there time taken *to prove* that there are, in fact, an infinite number of primes. There are many proofs to verify this belief. One such begins with the supposition that there are only a finite number of prime numbers, and then shows this to be false. For our purposes, we will say that there are only n primes, which we will refer to as $p_1, p_2, p_3, p_4, ..., p_n$. Suppose we let N equal the product of all of these n primes, that is, $N = p_1 \cdot p_2 \cdot p_3 \cdot p_4 \cdot ... \cdot p_n$. Clearly, $N + 1$ is greater than p_n and is not a prime number, since we listed all the primes earlier. (The only two primes that are consecutive numbers are 2 and 3.) Since it is not a prime number, it must have a divisor common with N, and we will call that divisor p_k, which is one of the primes listed above. Since p_k is a factor of both N and $N + 1$, it must also divide $(N + 1) - N = 1$, which is impossible. Therefore, the supposition that there is a finite number of primes has to be rejected, leaving us with the alternative that there are an infinite number of primes. This is a rather simple proof to justify something that most of us already have known from our school exposure to mathematics, but usually students are not exposed to a proof of this notion.

THE NEGLECTED TRIANGULAR NUMBERS

Throughout our schooling, among the general integers, we have heard about those that were square numbers, which we saw both arithmetically and geometrically. Arithmetically, square numbers were obtained by simply taking a natural number and multiplying it by itself. Geometrically, we saw square numbers as points that could be arranged in the form of the square, as shown in figure 1.5.

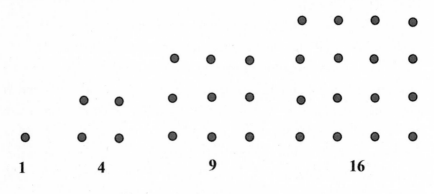

Figure 1.5

You will also recall that square numbers played a very important role in mathematics, in which one of the most famous equations is comprised entirely of square numbers. Here we refer to the well-known Pythagorean theorem, where we have $a^2 + b^2 = c^2$.

As mathematics is developed during the school years, square numbers are profuse but what hardly ever gets mentioned are *triangular numbers*. As you might imagine from their name, these are numbers that represent the number of dots that can be arranged as equilateral triangles, as shown in figure 1.6.

Figure 1.6

The following is a list of the triangular numbers less than 10,000:

1, 3, 6, 10, 15, 21, 28, 36, 45, 55, 66, 78, 91, 105, 120, 136, 153, 171, 190, 210, 231, 253, 276, 300, 325, 351, 378, 406, 435, 465, 496, 528, 561, 595, 630, 666, 703, 741, 780, 820, 861, 903, 946, 990, 1035, 1081, 1128, 1176, 1225, 1275, 1326, 1378, 1431, 1485, 1540, 1596, 1653, 1711, 1770, 1830, 1891, 1953, 2016, 2080, 2145, 2211, 2278, 2346, 2415, 2485, 2556, 2628, 2701, 2775, 2850, 2926, 3003, 3081, 3160, 3240, 3321, 3403, 3486, 3570, 3655, 3741, 3828, 3916, 4005, 4095, 4186, 4278, 4371, 4465, 4560, 4656, 4753, 4851, 4950, 5050, 5151, 5253, 5356, 5460, 5565, 5671, 5778, 5886, 5995, 6105, 6216, 6328, 6441, 6555, 6670, 6786, 6903, 7021, 7140, 7260, 7381, 7503, 7626, 7750, 7875, 8001, 8128, 8256, 8385, 8515, 8646, 8778, 8911, 9045, 9180, 9316, 9453, 9591, 9730, 9870.

Perhaps the most easily detectable property of triangular numbers (T_n) is that they each are a sum of the first n consecutive natural numbers, as you can see below from the first seven triangular numbers:

$$T_1 = 1$$
$$T_2 = 1 + 2 = 3$$
$$T_3 = 1 + 2 + 3 = 6$$
$$T_4 = 1 + 2 + 3 + 4 = 10$$

$$T_5 = 1 + 2 + 3 + 4 + 5 = 15$$
$$T_6 = 1 + 2 + 3 + 4 + 5 + 6 = 21$$
$$T_7 = 1 + 2 + 3 + 4 + 5 + 6 + 7 = 28$$

However, since these triangular numbers result from an arithmetic series, the formula to find the nth triangular number can be found with the following formula: $T_n = \dfrac{n(n+1)}{2}$, which we discussed earlier.

This is just the beginning of where the properties of triangular numbers take off. Let us now enjoy some of these truly unexpected properties of triangular numbers, the inclusion of which in our school instruction would certainly have given more "life" to the subject!

1. The sum of any two consecutive triangular numbers is equal to a square number, as see with these two examples:

$$T_1 + T_2 = 1 + 3 = 4 = 2^2$$
$$T_5 + T_6 = 15 + 21 = 36 = 6^2$$

2. By inspecting the above list of triangular numbers, you will notice that a triangular number never appears to end with the digits 2, 4, 7, or 9. This is true throughout all of the triangular numbers.

3. The number 3 is the only triangular number that is also a prime number. This can also be seen in our sample of triangular numbers provided earlier.

4. If 1 is added to 9 times a triangular number, the result will be another triangular number, such as in the case of the following example: $9 \cdot T_3 + 1 = 9 \cdot 6 + 1 = 54 + 1 = 55$, which is the tenth triangular number.

5. Analogous to the previous curiosity, we find that if 1 is added to 8 times a triangular number, the result will be a square number, as in the case for $8 \cdot T_3 + 1 = 8 \cdot 6 + 1 = 48 + 1 = 49 = 7^2$.

6. The sum of n consecutive cubes beginning with 1 equals the square of the nth triangular number. That is, $T_n^2 = 1^3 + 2^3 + 3^3 + 4^3 + \cdots + n^3$.

As an example, consider the following breakdown for the fifth triangular number: $T_5^2 = 1^3 + 2^3 + 3^3 + 4^3 + 5^3 = 1 + 8 + 27 + 64 + 125 = 225 = 15^2$.

7. Some triangular numbers are also palindromic numbers—numbers that read the same forward and backward. The first few of these are 1, 3, 6, 55, 66, 171, 595, 666, 3003, 5995, 8778, 15051, 66066, 617716, 828828, 1269621, 1680861, 3544453, 5073705, 5676765, 6295926, 351335153, 61477416, 178727871, 1264114621, 1634004361, and so on.

8. There are infinitely many triangular numbers that are also square numbers. The first few of these are $1 = 1^2$, $36 = 6^2$, $1225 = 35^2$, $41616 = 204^2$, $1413721 = 1189^2$, $48024900 = 6930^2$, $1631432881 = 40391^2$, and so on. Naturally, we can generate these square-triangular numbers with the following formula: $Q_n = 34Q_{n-1} - Q_{n-2} + 2$, where Q_n represents the nth square-triangular number. An interesting peculiarity among these square-triangular numbers is that all even square-triangular numbers are a multiple of 9.

9. There are some triangular numbers that produced another triangular number, when the order of the digits is reversed. The first few of these are: 1, 3, 6, 10, 55, 66, 120, 153, 171, 190, 300, 351, 595, 630, 666, 820, 3003, 5995, 8778, 15051, 17578, 66066, 87571, 156520, 180300, 185745, 547581, and so on.

10. Another curiosity involving triangular numbers is that certain members of the set of triangular numbers can be paired so that the sum and the difference of these pairs also results in a triangular number. Here are two examples of these triangular number pairs:

(15, 21) yields $21 - 15 = 6$ and $21 + 15 = 36$. Both 6 and 36 are also triangular numbers.

(105, 171) yields $171 - 105 = 66$ and $105 + 171 = 276$. Both 66 and 276 are also triangular numbers.

11. The properties of triangular numbers seem to be boundless. For example, there are only six triangular numbers that can be represented as the product of three consecutive numbers. These are:

$$T_3 = 6 = 1 \cdot 2 \cdot 3,$$
$$T_{15} = 120 = 4 \cdot 5 \cdot 6,$$
$$T_{20} = 210 = 9 \cdot 10 \cdot 11,$$
$$T_{44} = 990 = 9 \cdot 10 \cdot 11,$$
$$T_{608} = 185136 = 56 \cdot 57 \cdot 58,$$
$$T_{22736} = 258474216 = 636 \cdot 637 \cdot 638.$$

Of these, the number $T_{15} = 120$ is particularly "gifted" in that it can be represented as products of three, four, and five consecutive numbers. There has not been another such triangular number found that has this property. That is, $T_{15} = 120 = 4 \cdot 5 \cdot 6 = 2 \cdot 3 \cdot 4 \cdot 5 = 1 \cdot 2 \cdot 3 \cdot 4 \cdot 5$.

Furthermore, we can show that there are some triangular numbers that are the product of two consecutive numbers, as in the following examples:

$$T_3 = 6 = 2 \cdot 3,$$
$$T_{20} = 210 = 14 \cdot 15,$$
$$T_{119} = 7140 = 84 \cdot 85,$$
$$T_{696} = 242556 = 492 \cdot 493.$$

12. There are only six triangular numbers that are comprised of a unique digit. They are: 1, 3, 6, 55, 66, 666.

13. Among the Fibonacci numbers there are only four known triangular numbers: 1, 3, 21, and 55.

14. To write any positive integer as the sum of triangular numbers, you would not need more than three triangular numbers. For example, the first ten positive integers can be represented as the sum of triangular numbers as follows:

1,
2 = 1 + 1,
3,
4 = 1 + 3,
5 = 1 + 1 + 3,
6,
7 = 1 + 6,
8 = 1 + 1 + 6,
9 = 3 + 3 + 3, and
10 = 1 + 3 + 6.

15. Every fourth power of an integer greater than 1 is the sum of two triangular numbers. For example:

$$2^4 = 16 = T_3 + T_4 = T_1 + T_5$$
$$3^4 = 81 = T_8 + T_9 = T_5 + T_{11}$$
$$4^4 = 256 = T_{15} + T_{16} = T_{11} + T_{19}$$
$$5^4 = 625 = T_{24} + T_{25} = T_{19} + T_{29}$$
$$6^4 = 1296 = T_{35} + T_{36} = T_{29} + T_{41}$$
$$7^4 = 2401 = T_{48} + T_{49} = T_{41} + T_{55}$$

16. The sum of successive powers of 9 will result in a triangular number, as we can see from the following first few such examples:

$$1 = T_1$$
$$1 + 9 = T_4$$
$$1 + 9 + 9^2 = T_{13}$$
$$1 + 9 + 9^2 + 9^3 = T_{40}$$
$$1 + 9 + 9^2 + 9^3 + 9^4 = T_{121}$$

17. Another curious pattern with triangular numbers is shown below and should be obvious at first sight, looking at the subscripts:

$$T_1 + T_2 + T_3 = T_4$$
$$T_5 + T_6 + T_7 + T_8 = T_9 + T_{10}$$
$$T_{11} + T_{12} + T_{13} + T_{14} + T_{15} = T_{16} + T_{17} + T_{18}$$
$$T_{19} + T_{20} + T_{21} + T_{22} + T_{23} + T_{24} = T_{25} + T_{26} + T_{27} + T_{28}$$
$$T_{29} + T_{30} + T_{31} + T_{32} + T_{33} + T_{34} + T_{35} = T_{36} + T_{37} + T_{38} + T_{39} + T_{40}$$

Notice the symmetry and consistency in this arrangement.

In short, the introduction of triangular numbers, which seems to be missing from most school curricula, offers a delightful insight into the beauty of mathematics—one that seems to hold many opportunities for further exploration.

PERFECT NUMBERS

Although most mathematics teachers probably tell their students that everything in mathematics is considered to be perfect, there are in fact numbers that teachers and mathematicians alike have designated *perfect numbers*. The characteristic that these numbers all share is that they are numbers equal to the sum of their proper divisors, which are all the divisors except the number itself. The smallest perfect number is 6, since $6 = 1 + 2 + 3$, which is the sum of all its divisors excluding the number 6 itself. All perfect numbers, of which the first few are 6, 28, 496, 8,128, and so on, are also triangular numbers.

Before we explore the unusual characteristics of perfect numbers, we can have some fun by looking at some other peculiarities of the number 6. For example, it is the only number that is the sum and product of the same three numbers, since we also have $6 = 1 \cdot 2 \cdot 3$. Another curiosity of the number 6 can be seen with the following: $6 = \sqrt{1^3 + 2^3 + 3^3}$.

The next larger perfect number is 28, since once again we can show that it is equal to the sum of its proper divisors, or we can say the sum of its prime factors. Hence, $28 = 1 + 2 + 4 + 7 + 14$. We then have to go a long way to get to the next perfect number, which is 496, since again we have the sum of its proper divisors, namely, $496 = 1 + 2 + 4 + 8 + 16 + 31 + 62 + 124 + 248$. The first four perfect numbers (6, 28, 496, and 8,128) were known to the ancient Greeks. It was Euclid who is credited with establishing a theorem to enable us to find other perfect numbers. He said that for an integer k, if $2^k - 1$ is a prime number, then $2^{k-1}(2^k - 1)$ will give us a perfect number. We do not have to use all values of k, since if k is a composite number, then $2^k - 1$ is also a composite number.[4]

Using Euclid's method for generating perfect numbers, we get the following table (figure 1.7), in which, for the values of k, we get $2^{k-1}(2^k - 1)$ to be a perfect number, but only when $2^k - 1$ is a prime number.

Rank	k	Perfect Number	Number of Digits	Year Discovered
1	2	6	1	Known to the Greeks
2	3	28	2	Known to the Greeks
3	5	496	3	Known to the Greeks
4	7	8128	4	Known to the Greeks
5	13	33550336	8	1456
6	17	8589869056	10	1588
7	19	137438691328	12	1588
8	31	2305843008139952128	19	1772
9	61	265845599 . . . 953842176	37	1883
10	89	191561942 . . . 548169216	54	1911
11	107	131640364 . . . 783728128	65	1914
12	127	144740111 . . . 199152128	77	1876
13	521	235627234 . . . 555646976	314	1952
14	607	141053783 . . . 537328128	366	1952

15	1279	541625262 . . . 984291328	770	1952
16	2203	108925835 . . . 453782528	1327	1952
17	2281	994970543 . . . 139915776	1373	1952
18	3217	335708321 . . . 628525056	1937	1957
19	4253	182017490 . . . 133377536	2561	1961
20	4423	407672717 . . . 912534528	2663	1961
21	9689	114347317 . . . 429577216	5834	1963
22	9941	598885496 . . . 073496576	5985	1963
23	11213	395961321 . . . 691086336	6751	1963
24	19937	931144559 . . . 271942656	12003	1971
25	21701	100656497 . . . 141605376	13066	1978
26	23209	811537765 . . . 941666816	13973	1979
27	44497	365093519 . . . 031827456	26790	1979
28	86243	144145836 . . . 360406528	51924	1982
29	110503	136204582 . . . 603862528	66530	1988
30	132049	131451295 . . . 774550016	79502	1983
31	216091	278327459 . . . 840880128	130100	1985
32	756839	151616570 . . . 565731328	455663	1992
33	859433	838488226 . . . 416167936	517430	1994
34	1257787	849732889 . . . 118704128	757263	1996
35	1398269	331882354 . . . 723375616	841842	1996
36	2976221	194276425 . . . 174462976	1791864	1997
37	3021377	811686848 . . . 022457856	1819050	1998
38	6972593	955176030 . . . 123572736	4197919	1999
39	13466917	427764159 . . . 863021056	8107892	2001
40	20996011	793508909 . . . 206896128	12640858	2003
41	24036583	448233026 . . . 572950528	14471465	2004
42	25964951	746209841 . . . 791088128	15632458	2005
43	30402457	497437765 . . . 164704256	18304103	2005
44	32582657	775946855 . . . 577120256	19616714	2006
45	37156667	204534225 . . . 074480128	22370543	2008
46	42643801	144285057 . . . 377253376	25674127	2009
47	43112609	500767156 . . . 145378816	25956377	2008
48	57885161	169296395 . . . 270130176	34850340	2013

Figure 1.7

After the eighth perfect number, you will notice that our table was not able to accommodate the large size of the succeeding perfect numbers, so we present the ninth and tenth perfect numbers here to give the reader a sense of the magnitudes involved:

2,658,455,991,569,831,744,654,692,615,953,842,176

191,561,942,608,236,107,294,793,378,084,303,638,130,997,321,548,169,216

As we inspect our list of perfect numbers, we notice some of their properties. The terminal digits appear to always be either 6 or 28, and these are preceded by an odd digit. They also appear to be triangular numbers, which we recall are the sums of consecutive natural numbers such as:

$6 = 1 + 2 + 3$

$28 = 1 + 2 + 3 + 4 + 5 + 6 + 7$

$496 = 1 + 2 + 3 + 4 + \cdots + 28 + 29 + 30 + 31.$

There is another curious pattern among perfect numbers. We can see this by looking at every perfect number after 6; we notice that they can be represented as a partial sum of the series $1^3 + 3^3 + 5^3 + 7^3 + 9^3 + 11^3 + \cdots$. For the first few examples we have the following:

$28 = 1^3 + 3^3$

$496 = 1^3 + 3^3 + 5^3 + 7^3$

$8,128 = 1^3 + 3^3 + 5^3 + 7^3 + 9^3 + 11^3 + 13^3 + 15^3.$

An ambitious reader might try to show that the partial sums for the next perfect numbers follow this same pattern. To date, no odd perfect number has been found—even with the aid of computers. Yet, without a mathematical proof, we cannot rule out the possible existence of an odd perfect number.

MAKING MISTAKEN GENERALIZATIONS

It is easy to try to make a generalization when a pattern appears to be consistent. However, patterns can be consistent up to a certain point after which an inconsistency can disturb the pattern. This is something that is too often not discussed in the school curriculum. Perhaps teachers don't want to upset students by showing them that there can be such inconsistencies. Let's take a look at one such example: Can every odd number greater than 1 be expressed as the sum of a power of 2 and a prime number? Consider the listing below. As we begin with the smallest possible numbers and work our way along, we notice that through the number 51 the pattern remains in order. As we progress to the number 125, there is still no problem with the pattern remaining consistent. Then, as we try to see if this relationship holds for the number 127— much to our amazement—we find that the pattern no longer holds true. It then continues on and doesn't falter until the number 149.

$$3 = 2^0 + 2$$
$$5 = 2^1 + 3$$
$$7 = 2^2 + 3$$
$$9 = 2^2 + 5$$
$$11 = 2^3 + 3$$
$$13 = 2^3 + 5$$
$$15 = 2^3 + 7$$
$$17 = 2^2 + 13$$
$$19 = 2^4 + 3$$

.

.

.

$$51 = 2^5 + 19$$

.

.

.

$125 = 2^6 + 61$

$127 = ?$

$129 = 2^5 + 97$

$131 = 2^7 + 3$

This conjecture as a possible "rule" was originally proposed by the French mathematician Alphonse de Polignac (1817–1890), yet it falters for the following numbers: 127, 149, 251, 331, 337, 373, and 509. We now know that there are an infinite number of such failures of this conjecture, of which the number 2,999,999 is one.

Another illustration that appears to produce a delightful pattern but cannot be extended beyond a certain point is the following list of equalities that we find by taking certain numbers to powers of 1, 2, 3, 4, 5, 6, and 7:

$1^0 + 13^0 + 28^0 + 70^0 + 82^0 + 124^0 + 139^0 + 151^0 =$	$4^0 + 7^0 + 34^0 + 61^0 + 91^0 + 118^0 + 145^0 + 148^0$
$1^1 + 13^1 + 28^1 + 70^1 + 82^1 + 124^1 + 139^1 + 151^1 =$	$4^1 + 7^1 + 34^1 + 61^1 + 91^1 + 118^1 + 145^1 + 148^1$
$1^2 + 13^2 + 28^2 + 70^2 + 82^2 + 124^2 + 139^2 + 151^2 =$	$4^2 + 7^2 + 34^2 + 61^2 + 91^2 + 118^2 + 145^2 + 148^2$
$1^3 + 13^3 + 28^3 + 70^3 + 82^3 + 124^3 + 139^3 + 151^3 =$	$4^3 + 7^3 + 34^3 + 61^3 + 91^3 + 118^3 + 145^3 + 148^3$
$1^4 + 13^4 + 28^4 + 70^4 + 82^4 + 124^4 + 139^4 + 151^4 =$	$4^4 + 7^4 + 34^4 + 61^4 + 91^4 + 118^4 + 145^4 + 148^4$
$1^5 + 13^5 + 28^5 + 70^5 + 82^5 + 124^5 + 139^5 + 151^5 =$	$4^5 + 7^5 + 34^5 + 61^5 + 91^5 + 118^5 + 145^5 + 148^5$
$1^6 + 13^6 + 28^6 + 70^6 + 82^6 + 124^6 + 139^6 + 151^6 =$	$4^6 + 7^6 + 34^6 + 61^6 + 91^6 + 118^6 + 145^6 + 148^6$
$1^7 + 13^7 + 28^7 + 70^7 + 82^7 + 124^7 + 139^7 + 151^7 =$	$4^7 + 7^7 + 34^7 + 61^7 + 91^7 + 118^7 + 145^7 + 148^7$

From these examples one could easily form the conclusion that for the natural number n the following should hold:

$$1^n + 13^n + 28^n + 70^n + 82^n + 124^n + 139^n + 151^n = 4^n + 7^n + 34^n + 61^n + 91^n + 118^n + 145^n + 148^n$$

These values are shown in figure 1.8.

n	Sums
0	8
1	608
2	70,076
3	8,953,712
4	1,199,473,412
5	165,113,501,168
6	23,123,818,467,476
7	3,276,429,220,606,352

Figure 1.8

To make a generalization of this pattern would be a predictable behavior. However, at the same time it would also be a marvelous mistake. This mistake does not manifest itself until we take the next case, $n = 8$. We notice that the two sums are no longer the same:

$1^8 + 13^8 + 28^8 + 70^8 + 82^8 + 124^8 + 139^8 + 151^8 =$
468,150,771,944,932,292
$4^8 + 7^8 + 34^8 + 61^8 + 91^8 + 118^8 + 145^8 + 148^8 =$
468,087,218,970,647,492

As a matter of fact, the difference between these two sums is

$468,150,771,944,932,292 - 468,087,218,970,647,492 =$
63,552,974,284,800.

As n increases so does the difference between the two sums. For $n = 20$, the difference between the sums is

3,388,331,687,715,737,094,794,416,650,060,343,026,048,000.

This illustration shows how important it is to avoid such mistakes by proving what appears to be a generalization, rather than merely assuming that a pattern is established from a few early examples.

THE FIBONACCI NUMBERS

Perhaps the most ubiquitous set of numbers in all of mathematics is the *Fibonacci numbers*. These numbers relate to almost every aspect of mathematics that is presented in the school curriculum. Yet, for some mysterious reason, they are typically not part of the instructional program. A resourceful teacher may find creative ways to deviate from the prescribed curriculum and introduce these wonderful numbers to students who could then be easily enchanted by mathematics.

These numbers stem from a problem on the regeneration of rabbits found in chapter 12 of a book titled *Liber Abaci*, which was published by Leonardo of Pisa, known today as Fibonacci, in the year 1202. The problem was to determine how many rabbits there would be at the end of one year following the prescribed scheme of the problem. When you list the number of rabbits that exist each month, the following sequence evolves: 1, 1, 2, 3, 5, 8, 13, 21, 34, 55, 89, 144. A quick inspection of the sequence shows that, beginning with the first two numbers, each successive number is the sum of the two previous ones.

Now you may be wondering what makes this sequence of numbers so spectacular. One spectacular aspect of the sequence is that it amazingly relates to the *golden ratio*, which we present in chapter 3. If you take successive ratios of consecutive numbers in the Fibonacci sequence, their value will get ever closer to the golden ratio, as we show in the following list (figure 1.9), where F_n represents the nth Fibonacci number:

The Ratios of Consecutive Fibonacci Numbers[5]

$\dfrac{F_{n+1}}{F_n}$	$\dfrac{F_n}{F_{n+1}}$
$\dfrac{1}{1} = 1.000000000$	$\dfrac{1}{1} = 1.000000000$
$\dfrac{2}{1} = 2.000000000$	$\dfrac{1}{2} = 0.500000000$
$\dfrac{3}{2} = 1.500000000$	$\dfrac{2}{3} = 0.666666667$
$\dfrac{5}{3} = 1.666666667$	$\dfrac{3}{5} = 0.600000000$
$\dfrac{8}{5} = 1.600000000$	$\dfrac{5}{8} = 0.625000000$
$\dfrac{13}{8} = 1.625000000$	$\dfrac{8}{13} = 0.615384615$
$\dfrac{21}{13} = 1.615384615$	$\dfrac{13}{21} = 0.619047619$
$\dfrac{34}{21} = 1.619047619$	$\dfrac{21}{34} = 0.617647059$
$\dfrac{55}{34} = 1.617647059$	$\dfrac{34}{55} = 0.618181818$
$\dfrac{89}{55} = 1.618181818$	$\dfrac{55}{89} = 0.617977528$
$\dfrac{144}{89} = 1.617977528$	$\dfrac{89}{144} = 0.618055556$
$\dfrac{233}{144} = 1.618055556$	$\dfrac{144}{233} = 0.618025751$
$\dfrac{377}{233} = 0.618025751$	$\dfrac{233}{377} = 0.618037135$
$\dfrac{610}{377} = 1.618037135$	$\dfrac{377}{610} = 0.618032787$
$\dfrac{987}{610} = 1.618032787$	$\dfrac{610}{987} = 0.618034448$

Figure 1.9

This is one way that the Fibonacci numbers—whose ratio of consecutive members approaches the golden ratio—can relate to art and architecture, and yet we can also show that they relate to biology. For example, if you count the spirals on a pineapple, you will find that in one direction there are 8 spirals going around, and in the other direction there are two kinds of spirals: one having 5 spirals and the other having 13 spirals. In other words, the numbers of spirals on a pineapple are represented by the Fibonacci numbers 5, 8, and 13. Typical pinecones also have spirals, in one direction there are 8 spirals; in the other direction, there are 13 spirals.

There are boundless curiosities about the Fibonacci numbers waiting to be discovered by eager students of mathematics. In 1963, the Fibonacci Association was founded to enable mathematicians to share new findings regarding the Fibonacci numbers through the *Fibonacci Quarterly*, which is still in publication today. For example, one would think that the Fibonacci numbers would have nothing in common with the Pythagorean theorem. Well, here's a surprise. Follow this technique, and you will see how you can generate Pythagorean triples—that is, three numbers that satisfy the equation $a^2 + b^2 = c^2$—from any four consecutive numbers of the Fibonacci sequence.

To make a Pythagorean triple from the Fibonacci numbers, we take any four consecutive numbers in the Fibonacci sequence, such as 3, 5, 8, and 13. We now follow these rules:

1. Multiply the middle two numbers and double the result.

    Here the product of 5 and 8 is 40; then we double this to get **80**, which will be one member of the Pythagorean triple.

2. Multiply the two outer numbers.

    Here the product of 3 and 13 is **39**, which will be another member of the Pythagorean triple.

3. Add the squares of the inner two numbers to get the third member of the Pythagorean triple.

Here $5^2 + 8^2 = 25 + 64 = \mathbf{89}$.

So we have found a Pythagorean triple: (39, 80, 89). We can verify that this is, in fact, a Pythagorean triple by showing that $39^2 + 80^2 = 1{,}521 + 6{,}400 = 7{,}921 = 89^2$.

We offer here a mere sampling of some of other startling relationships involving the Fibonacci numbers.

1. The sum of any ten consecutive Fibonacci numbers is divisible by 11.

   $$11 \mid (F_n + F_{n+1} + F_{n+2} + \ldots + F_{n+8} + F_{n+9}).$$

   For example: $5 + 8 + 13 + 21 + 34 + 55 + 89 + 144 + 233 + 377 = 979$, which is $89 \cdot 11$.

2. Any two consecutive Fibonacci numbers are relatively prime. That is, their greatest common divisor is 1.

3. Fibonacci numbers in a composite number position—for example, those Fibonacci numbers in the nonprime positions (with the exception of the fourth Fibonacci number)—are also composite (nonprime) numbers. Another way of saying this is that if n is not prime, then F_n is not prime, where $n \neq 4$, since the one exception is $F_4 = 3$, which is a prime number.

4. The sum of the first n Fibonacci numbers is equal to the $(n + 2)$ Fibonacci number minus 1. This can be written as

   $$\sum_{i=1}^{n} F_i = F_1 + F_2 + F_3 + F_4 + \ldots + F_n = F_{n+2} - 1.$$

   For example, the sum of the first nine Fibonacci numbers is

   $$1 + 1 + 2 + 3 + 5 + 8 + 13 + 21 + 34 = 88 = 89 - 1.$$

5. The sum of the first n consecutive even-positioned Fibonacci numbers is 1 less than the Fibonacci number that follows the last even number in the sum, which is written symbolically as

$$\sum_{i=1}^{n} F_{2i} = F_2 + F_4 + F_6 + \ldots + F_{2n-2} + F_{2n} = F_{2n+1} - 1.$$

For example, $1 + 3 + 8 + 21 + 55 + 144 = 232 = 233 - 1$.

6. The sum of the first n consecutive odd-positioned Fibonacci numbers is equal to the Fibonacci number that follows the last odd number in the sum, which is written symbolically as

$$\sum_{i=1}^{n} F_{2i-1} = F_1 + F_3 + F_5 + \ldots + F_{2n-3} + F_{2n-1} = F_{2n}.$$

For example, $1 + 2 + 5 + 13 + 34 + 89 = 144$.

7. The sum of the squares of the initial Fibonacci numbers is equal to the product of the last number in the sequence and its successor; written symbolically this is

$$\sum_{i=1}^{n} \left(F_i \right)^2 = F_n F_{n+1}.$$

For example, $1^2 + 1^2 + 2^2 + 3^2 + 5^2 + 8^2 + 13^2 = 273 = 13 \cdot 21$.

8. The difference of the squares of two alternate Fibonacci numbers (that is, two Fibonacci numbers that are separated by one other member of the sequence) is equal to the Fibonacci number in the sequence whose position number is the sum of their position numbers. We write this symbolically as $F_n^2 - F_{n+2}^2 = F_{2n-2}$.

For example, $13^2 - 5^2 = 169 - 25 = 144$, which is the 7th + 5th = 12th Fibonacci number.

9. The sum of the squares of two consecutive Fibonacci numbers is equal to the Fibonacci number in the sequence whose position number is the sum of their position numbers; written symbolically, this is $F_n^2 + F_{n+1}^2 = F_{2n+1}$.

For example, $8^2 + 13^2 = 233$, which is the 6th + 7th = 13th Fibonacci number.

10. For any group of four consecutive Fibonacci numbers, the difference of the squares of the middle two numbers is equal to the product of the outer two numbers. Symbolically, we can write this as $F_{n+1}^2 - F_n^2 = F_{n-1} \cdot F_{n+2}$.

   Consider the four consecutive Fibonacci numbers 3, 5, 8, 13. The following then holds true: $8^2 - 5^2 = 39 = 3 \cdot 13$.

11. The product of two alternate Fibonacci numbers is 1 more or less than the square of the Fibonacci number between them. Symbolically, we can write this as $F_{n-1} \cdot F_{n+1} = F_n^2 + (-1)^n$. If n is even, the product is 1 more; if n is odd, the product is 1 less.

   This can be extended to the following: The difference between the square of the selected Fibonacci number and the various products of Fibonacci numbers equidistant from the selected Fibonacci number is the square of another Fibonacci number:

$$F_{n-k}F_{n+k} - F_n^2 = \pm F_k^2, \quad \text{where } n \geq 1, \text{ and } k \geq 1.$$

12. A Fibonacci number F_{mn} is divisible by a Fibonacci number F_m. (We can write this as: $F_m \mid F_{mn}$, which reads "F_m divides F_{mn}.") Another way of looking at this is: if p is divisible by q, then F_p is divisible by F_q. In symbols, $q \mid p \Rightarrow F_q \mid F_p$, where, m, n, p, and q are positive integers.

   Here is how this looks for specific cases:

$F_1 \mid F_n$, i.e., $1 \mid F_1, 1 \mid F_2, 1 \mid F_3, 1 \mid F_4, 1 \mid F_5, 1 \mid F_6, \ldots, 1 \mid F_n, \ldots$

$F_2 \mid F_{2n}$, i.e., $1 \mid F_2, 1 \mid F_4, 1 \mid F_6, 1 \mid F_8, 1 \mid F_{10}, 1 \mid F_{12}, \ldots, 1 \mid F_{2n}, \ldots$

$F_3 \mid F_{3n}$, i.e., $2 \mid F_3, 2 \mid F_6, 2 \mid F_9, 2 \mid F_{12}, 2 \mid F_{15}, 2 \mid F_{18}, \ldots, 2 \mid F_{3n}, \ldots$

$F_4 \mid F_{4n}$, i.e., $3 \mid F_4, 3 \mid F_8, 3 \mid F_{12}, 3 \mid F_{16}, 3 \mid F_{20}, 3 \mid F_{24}, \ldots, 3 \mid F_{4n}, \ldots$

$F_5 \mid F_{5n}$, i.e., $5 \mid F_5, 5 \mid F_{10}, 5 \mid F_{15}, 5 \mid F_{20}, 5 \mid F_{25}, 5 \mid F_{30}, \ldots, 5 \mid F_{5n}, \ldots$

$F_6 \mid F_{6n}$, i.e., $8 \mid F_6, 8 \mid F_{12}, 8 \mid F_{18}, 8 \mid F_{24}, 8 \mid F_{30}, 8 \mid F_{36}, \ldots, 8 \mid F_{6n}, \ldots$

$F_7 \mid F_{7n}$, i.e., $13 \mid F_7, 13 \mid F_{14}, 13 \mid F_{21}, 13 \mid F_{28}, 13 \mid F_{35}, 13 \mid F_{42}, \ldots, 13 \mid F_{7n}, \ldots$

13. The sum of the products of consecutive Fibonacci numbers is either the square of a Fibonacci number or 1 less than the square of a Fibonacci number; shown symbolically, this is

$$\sum_{i=2}^{n+1} F_i F_{i-1} = F_{n+1}^2, \quad \text{when } n \text{ is odd.}$$

$$\sum_{i=2}^{n+1} F_i F_{i-1} = F_{n+1}^2 - 1, \quad \text{when } n \text{ is even.}$$

We hope that this very brief introduction to this most important and omnipresent sequence of numbers will motivate you to consider the myriad of other sightings and applications. If you wish to explore Fibonacci numbers in greater depth, one comprehensive source is *The Fabulous Fibonacci Numbers*, by A. S. Posamentier and I. Lehmann (Amherst, NY: Prometheus Books, 2007).

CHAPTER 2

ALGEBRAIC EXPLANATIONS OF ACCEPTED CONCEPTS

When you think back to your school days and specifically about what you learned in algebra, you typically see this as a series of mechanical procedures or algorithms leading to a desired result. Yet algebra has so much more to offer that unfortunately does not often find its way into the school curriculum. It allows us to understand verbal problems more easily. It enables us to explain mathematical concepts, which oftentimes can be seen as proofs, or simply justifications. In any case, if more of these kinds of algebraic uses were put into the school curriculum, more joy and pleasure could be derived from mathematics. In this chapter, we will show how certain ideas can be explained, and how certain concepts can become more meaningful, through the use of algebra. We will also show some algebraic techniques that most likely were not presented as part of your school learning, yet can be quite useful, both in mathematics and beyond.

SIMPLE ALGEBRA HELPS LOGICAL REASONING

Unfortunately, our training leads us to believe that algebra is merely a mechanical process that is needed to pursue higher mathematics. However, there are times when a rather perplexing reasoning problem can be best sorted out with very simple algebra. Consider the following problem: You are seated at a table in a dark room. On the table, there are twelve pennies, five of which are heads up and seven of which are tails up. Now mix the coins and separate them into two piles of five and seven coins, respectively. Because you are in a dark room, you will not know if the coins you are touching were heads up or tails up. Then flip over the coins in the five-coin pile. When the lights are turned on, there will be an equal number of heads in each of the two piles. How can this be possible?

Your first reaction is "you must be kidding!" "How can anyone do this task without seeing which coins are heads up or tails up?" You might actually want to try it with twelve coins to see if this really is true. To analyze the situation, a most clever (yet incredibly simple) use of algebra will be the key to the solution. Here is what we can do. Let's run through the problem again: First, set twelve coins on a table, such that there are five heads up and seven tails up. Then randomly (without seeing the heads or tails) select five coins for one pile and seven coins for another pile. When you separate the coins in the dark room, h heads will end up in the seven-coin pile. Then the other pile, the five-coin pile, will have $5 - h$ heads and $5 - (5 - h)$ tails. When you flip all the coins in the smaller pile, the $5 - h$ heads become tails and the $5 - (5 - h)$ tails become heads. Now each pile contains h heads!

As you can see from this one example, algebra can help with problem solving—and it can be fun and surprising, too!

DIVISION BY ZERO

Your math teachers have undoubtedly told you that dividing by zero is illegal—that is, dividing by zero is "undefined," and mathematics can stay consistent as long as we avoid this action. But why? Why is dividing by zero such a crime in mathematics? Consider the following expression: $a + b = c$. Let's do some algebra and see what happens. Subtract the c on both sides: $a + b - c = 0$.

Now, multiply both sides by 3, noticing that anything times zero is zero: $3(a + b - c) = 3 \cdot 0 = 0$.

Instead of using 3, we could also have done this for another number, say, 4: $4(a + b - c) = 4 \cdot 0 = 0$.

Okay, so far, so good. Let's write an easy identity: $0 = 0$.

Since both of these expressions are equal to 0, we can equate them as $3(a + b - c) = 4(a + b - c)$.

Both sides share the same factor $a + b - c$ so let's simplify here and divide both sides by this common factor: $\dfrac{3(a+b-c)}{a+b-c} = \dfrac{4(a+b-c)}{a+b-c}$. This leaves us with $3 = 4$, which is an absurdity!

What just happened? In fact, the same steps can be used to show that any number equals any other number, not just $3 = 4$. Of course, this is all wrong! Where did we err? We divided both sides of the equation by $a + b - c$. Notice that $a + b - c = 0$, hence, we divided by zero. It follows that if dividing by zero can lead to nonsense like $3 = 4$, then there is good reason why this division is forbidden.

Another example that can be used to emphasize this point is the following:

Begin by letting $a = b$.

Then multiply both sides of this equality by a to get $a^2 = ab$.

Then subtract b^2 from both sides of this equality to get
$a^2 - b^2 = ab - b^2$.

This can be factored as follows: $(a + b)(a - b) = b(a - b)$.

By dividing both sides by $(a - b)$, we get $a + b = b$.

However, since $a = b$, by substitution, we can get $2b = b$.

If we now divide both sides by b, we get the ridiculous result that
$2 = 1$.

Perhaps by now you will see why this happened. At the point that we divided both sides of the equality by $(a - b)$, we actually divided by zero, since $a = b$. Here you have further evidence why dividing by zero is outlawed in mathematics!

THE IRRATIONALITY OF THE SQUARE ROOT OF 2

We all know that there exist numbers that are not rational, that is, numbers that cannot be written as the quotient of two integers. Such numbers are called *irrational numbers*. The square root of 2 is probably the most popular example for an irrational number; other famous examples are $\pi = 3.1415926\ldots$ and Euler's number, $e = 2.7182818\ldots$. But how do we know that these numbers are irrational? The irrationality of $\sqrt{2}$ was already discovered by the Pythagoreans and, according to legend, they wanted to keep this discovery secret. It didn't remain a secret for long and, as the legend goes, Hippasus of Metapontum (ca. the fifth century BCE), one of the Pythagoreans, was drowned at sea as a punishment for divulging this secret knowledge.[1]

There is a very clever line of reasoning showing that $\sqrt{2}$ cannot be rational. It does not require anything from higher mathematics, but it has a logical twist that makes it not so easy to grasp when you see it for the first time. A mathematician would consider this kind of reasoning a proof by contradiction. The trick of proving that $\sqrt{2}$ is not rational is to assume that it would be rational and then use deductive reasoning to show that this assumption leads to a logical contradiction. Since the assumption of $\sqrt{2}$ being rational leads to a contradiction, it must be false, and, therefore, $\sqrt{2}$ cannot be rational.

Here is the proof. Assume that $\sqrt{2} = \dfrac{p}{q}$, where $\dfrac{p}{q}$ is an irreducible fraction, or a fraction reduced to lowest terms. Multiplying by q and taking the square on both sides of the equation leads to $2q^2 = p^2$. This implies that p^2 is an even number. But if p^2 is even, then p must be even as well (since if p were odd, then p^2 would also be odd, as can be easily seen). So we can represent p as an even number by representing it as $p = 2k$ for some positive integer k. Substituting this value of p into the earlier equation yields $2q^2 = 4k^2$, and thus $q^2 = 2k^2$. Now this implies that q^2 is also even, and we can simply repeat the argument from before to conclude that q must be even as well. But if both p and q

are even, then $\frac{p}{q}$ is not an irreducible fraction! This, however, contradicts our assumption that $\sqrt{2} = \frac{p}{q}$, where $\frac{p}{q}$ is an irreducible fraction. Since this assumption leads to a contradiction, logic tells us that our original assumption must be false. Hence $\sqrt{2}$ cannot be represented as an irreducible fraction and is, therefore, not rational—or, as we call it, *irrational*.

You might want to review the steps of this proof to better comprehend it, but once you have, you will enjoy its simplicity and appreciate the power of logical reasoning.

There also exists a "geometric proof" of the irrationality of $\sqrt{2}$ that was discovered by the American mathematician Stanley Tennenbaum (1927–2005). Its line of reasoning is similar to the one in the proof presented above, however, it offers a different perspective.

This proof starts completely analogously, by assuming that there exist smallest positive integers p and q such that $\sqrt{2} = \frac{p}{q}$ and therefore, $2q^2 = p^2$. In geometric terms, the equation $2q^2 = p^2$ means that the area of a square with side p is exactly twice the area of a square with side q and, by our assumption, there are no smaller squares with integer side lengths for which this is true. Now we put two squares of side q into a square of side p, as we show it in figure 2.1.

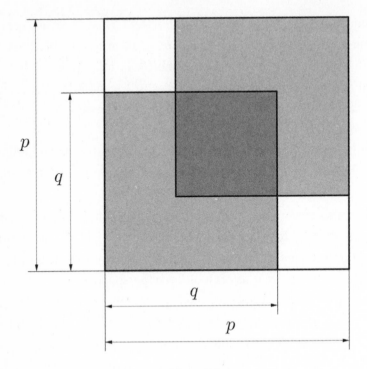

Figure 2.1

If the sum of the areas of the two shaded squares of side q is exactly equal to the area of the large square of side p, then the overlapping region of the two shaded squares, forming the darker center square, must be exactly equal to the sum of the areas of the two small white squares. But since p and q are integers, both the center square and the small white squares must have integer side lengths as well (the center square has sides of length $p - 2(p - q) = 2q - p$ and the small white square has sides of length $p - q$). Therefore, the center square and the two white squares represent smaller squares than the original ones, where again one has exactly twice the area of the other, contradicting our assumption that there exist smallest squares of integer side length with this property. Let's now look at how we can find the approximate value of an irrational number shown in square-root form.

BISECTION METHOD TO APPROXIMATE SQUARE ROOTS

The irrational numbers have nonterminating decimal expansions with no periodic repeating pattern among the digits. Many square roots in particular are irrational. How do we approximate a square root such as $\sqrt{2}$ with a finite decimal expansion? The *bisection method* provides an answer to this question.

Consider the graph of $y = x^2 - 2$, which is shown in figure 2.2.

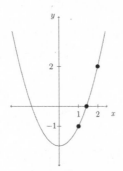

Figure 2.2

Notice that the graph intersects the x-axis at two points. When you solve $x^2 - 2 = 0$, you get the two solutions $x = \pm\sqrt{2}$. The positive root is the one we are interested in, and it is represented by the x-intercept to the right of the y-axis.

The graph helps us see why the positive $\sqrt{2}$ actually lies between 1 and 2. If we plug in $x = 1$, we get $y = 1^2 - 2 = -1$, so $(1,-1)$ is a point on the graph. If we plug in $x = 2$, we get $y = 2^2 - 2 = 2$, so $(2,2)$ is a point on the graph. The graph of the parabola $y = x^2 - 2$ is continuous, that is, you can draw the graph without lifting your pen. As such, going from left to right, from $(1, -1)$ to $(2, 2)$, the graph passes from the negative y-coordinate $y = -1$ (at $x = 1$) to the positive y-coordinate $y = 2$ (at $x = 2$) by crossing through the y-coordinate 0 at the x-intercept where $\sqrt{2}$ is located. Along this journey, the x-coordinates increase from 1 to $\sqrt{2}$ to 2, hence, we have the inequality $1 < \sqrt{2} < 2$.

The previous discussion leads to an algorithm to get better and better approximations of $\sqrt{2}$. Consider the interval [1, 2]. We can approximate $\sqrt{2}$ with the midpoint of this interval, hence, $\sqrt{2} \approx 1.5$. In order to improve the approximation, we will work with a smaller interval and use the midpoint of the smaller interval to approximate $\sqrt{2}$. We begin by computing the corresponding y-coordinate for $x = 1.5$, which is $y = (1.5)^2 - 2 = 0.25$.

Notice that this is a positive number.

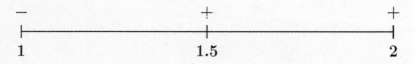

Figure 2.3

In figure 2.3 we represent the interval [1, 2] and include plus and minus signs to denote whether the corresponding y-coordinates for the x-coordinates shown are positive or negative. Cutting this interval in half, we know that $\sqrt{2}$ lies in either the left half [1, 1.5] or the right half [1.5, 2]. The sign change in the y-coordinates dictates that $\sqrt{2}$ will be in [1, 1.5] for the reasons discussed earlier. Notice that cutting the interval in half to narrow down the location of $\sqrt{2}$ is what gives rise to the name "bisection" method. The midpoint of [1, 1.5] is 1.25, hence we have a new approximation $\sqrt{2} \approx 1.25$ at this step of the process. (See figure 2.4.)

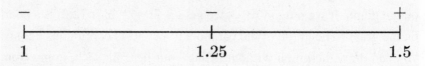

Figure 2.4

For the interval [1, 1.5], calculating $y = (1.25)^2 - 2 = -0.4375$ gives us the minus sign above $x = 1.25$. The change in signs occurs for the right half [1.25, 1.5].

The midpoint of [1.25, 1.5] is $\frac{1.25 + 1.5}{2} = 1.375$. Hence, we update our approximation $\sqrt{2} \approx 1.375$ at this step. (See figure 2.5.)

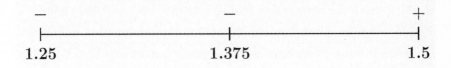

Figure 2.5

For the interval [1.25, 1.5], calculating $y = (1.375)^2 - 2 = -0.109375$ gives us the minus sign above $x = 1.375$. The change in signs occurs for the right half [1.375, 1.5].

The midpoint of [1.375, 1.5] is $\frac{1.375 + 1.5}{2} = 1.4375$. Hence, we have $\sqrt{2} \approx 1.4375$ at this step. It is possible to continue this process to get better and better approximations, but let us stop at this stage with the final approximation $\sqrt{2} \approx 1.4375$. How accurate is this approximation?

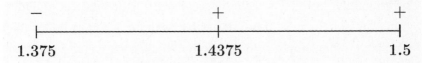

Figure 2.6

Consider figure 2.6 and notice that $\sqrt{2}$ actually lies in the left half [1.375, 1.4375]. However, let's pretend that we didn't know this fact. In this final stage, there are two potential choices for the location of $\sqrt{2}$, either the left half, [1.375, 1.4375], or the right half, [1.4375, 1.5].

The absolute value of the error in using 1.4375 as our approximation for $\sqrt{2}$ is represented by the length of the line segment connecting $\sqrt{2}$ with 1.4375, that is, $|\sqrt{2} - 1.4375|$. We use absolute value bars here to guarantee that we are dealing with the positive distance of the line segment joining $\sqrt{2}$ and 1.4375, regardless of whether $\sqrt{2}$ is to the left or right of 1.4375.

This line segment representing the absolute value of the error is inside either the left half or the right half of [1.375, 1.5]. Both the left half and right half have the same length, that is, 1.4375 – 1.375 = 0.0625 and 1.5 – 1.4375 = 0.0625. Thus, this line segment has length less than 0.0625 and the absolute value of the error is consequently bounded by 0.0625. Notice that by using the actual value of $\sqrt{2}$ = 1.4142..., we have the actual error 1.4375 – $\sqrt{2}$ ≈ 0.0233, which is less than our error bound 0.0625.

By using smaller and smaller intervals, we have smaller and smaller error bounds, so the bisection method can approximate $\sqrt{2}$, or other square roots, to as high an accuracy as we desire, provided we go through a sufficient number of steps. Although it is not the most efficient technique to find such approximations, the beauty of the bisection method lies in its simplicity. The basic arithmetic used in the bisection method is enough to get precise approximations of square roots, or the roots of other continuous functions in general. Unfortunately, in today's technological world students are guided to extract the square root of a number by pushing a button rather than by understanding how to do so manually. That is, we can get some appreciation as to what the computer is actually doing for you in this process of pressing a button. Irrational numbers can also be expressed as infinite continued fractions, as we will see in the next section.

CONTINUED FRACTIONS OF SQUARE ROOTS

Square roots may seem a bit mysterious because they are irrational and hence lack a periodic repeating pattern in their infinite decimal expansions. An example of this would be the number we just considered,

$\sqrt{2} = 1.414213562 \ldots$ The mystery lies in what digit comes next if there is no repeating pattern. However, if we are willing to go beyond decimal expansions and work with continued fractions, a topic in number theory, then we can see a beautiful repeating pattern for square roots. For example, when written as an infinite continued fraction, $\sqrt{2}$ looks like this:

$$\sqrt{2} = 1 + \cfrac{1}{2 + \cfrac{1}{2 + \cfrac{1}{2 + \cfrac{1}{2 + \cfrac{1}{2 + \ddots}}}}}$$

This pattern continues forever, hence the name *infinite continued fraction*. The repetition is in stark contrast with the much more irregular decimal expansion of $\sqrt{2}$. Why does this continued fraction equal $\sqrt{2}$? Let's dive into the algebra for an explanation.

Starting with the difference of two squares, we get:

$$\left(\sqrt{2} - 1\right)\left(\sqrt{2} + 1\right) = \left(\sqrt{2}\right)^2 - 1^2$$
$$\left(\sqrt{2} - 1\right)\left(\sqrt{2} + 1\right) = 2 - 1 = 1$$

Next, dividing both sides of this equality by $\sqrt{2} + 1$, we get:

$$\frac{\left(\sqrt{2} - 1\right)\left(\sqrt{2} + 1\right)}{\sqrt{2} + 1} = \frac{1}{\sqrt{2} + 1}$$
$$\sqrt{2} - 1 = \frac{1}{\sqrt{2} + 1}$$
$$\sqrt{2} = 1 + \frac{1}{\sqrt{2} + 1}, \text{ or } \quad \sqrt{2} = 1 + \frac{1}{1 + \sqrt{2}}$$

We will take the $\sqrt{2}$ in the denominator on the right side (i.e., the $\sqrt{2}$ appearing in $1 + \dfrac{1}{1+\sqrt{2}}$) and substitute it with the entire expression

$$\sqrt{2} = 1 + \frac{1}{1+\sqrt{2}} \text{ so that we get:}$$

$$\sqrt{2} = 1 + \cfrac{1}{1 + \ 1 + \cfrac{1}{1+\sqrt{2}}}$$

$$\sqrt{2} = 1 + \cfrac{1}{2 + \cfrac{1}{1+\sqrt{2}}}$$

At this point, we once again substitute the $\sqrt{2}$ appearing in the farthest denominator on the right side with $\sqrt{2} = 1 + \dfrac{1}{1+\sqrt{2}}$:

$$\sqrt{2} = 1 + \cfrac{1}{2 + \cfrac{1}{1 + \ 1 + \cfrac{1}{1+\sqrt{2}}}}$$

$$\sqrt{2} = 1 + \cfrac{1}{2 + \cfrac{1}{2 + \cfrac{1}{1+\sqrt{2}}}}$$

It is clear that this substitution of $\sqrt{2} = 1 + \dfrac{1}{1+\sqrt{2}}$ into the right side can continue forever, hence we have the earlier infinite continued fraction expansion of $\sqrt{2}$.

In general, the continued fraction expansions of irrational square roots will exhibit some repeating pattern in the denominators, revealing a structure to square roots that is hidden in their decimal expansions. Furthermore, continued fractions can also be used to obtain accurate rational approximations of square roots, that is, approximate square roots as fractions. As such, continued fractions lift a significant portion of the mystery surrounding irrational numbers like square roots.

FERMAT'S METHOD OF FACTORING

How do we know if an integer $n > 2$ is prime or composite? Many will recall the "square root test," which is to check to see if any of the primes less than or equal to $\sqrt{n}$ divides perfectly into n itself. If any such primes divide into n, then n is composite. Otherwise, n is prime.

However, the square root test is not always the most efficient way to determine whether a number is prime or composite. For example, consider testing whether $n = 6{,}499$ is prime or composite by using the square root test. Notice that $\sqrt{6499} \approx 80.616$ and there are quite a few primes less than 80.616, which we would have to test for divisibility.

Another way to determine whether n is prime or composite is to use Fermat's method of factoring. A numerical example will make this algorithm clear, so let's try $n = 6{,}499$. Start by setting x to be the smallest integer greater than or equal to $\sqrt{6499} \approx 80.616$. Because $\sqrt{6499}$ is not an integer, we will start with $x = 81$.

Compute $x^2 - n$, and see if the result is a perfect square: $81^2 - 6499 = 62$. Since 62 is not a perfect square, we increase x and set $x = 82$. Repeat this process of increasing x until we get a perfect square value for $x^2 - n$, which will happen with $x = 82$, since $82^2 - 6499 = 225 = 15^2$ At this point, set y to be the square root of the perfect square value of $x^2 - n$. In this example, $y = 15$. Fermat's method of factoring gives us $n = (x + y)(x - y)$:

6,499 = (82 + 15)(82 − 15)
6,499 = 97 · 67

This nontrivial factoring demonstrates 6,499 is indeed composite. Fermat's method of factoring is based on the familiar difference-of-two-squares formula from algebra: $x^2 - y^2 = (x + y)(x - y)$. The main idea is to write n as a difference of two squares, $n = x^2 - y^2$, with $x > y$, in order to make use of the difference of two-squares factoring. Rearranging, we get $x^2 - n = y^2$.

The goal is to systematically try various x values in $x^2 - n$ and see if this expression becomes a perfect square, which we will call y^2. This yields the factoring $n = (x + y)(x - y)$. Notice that $x^2 - n \geq 0$ implies $x \geq \sqrt{n}$, assuming x is positive. Thus, we can start with x being the smallest integer greater than or equal to $\sqrt{n}$ when substituting x values into $x^2 - n$.

This algorithm will always stop. The algorithm ends with either a nontrivial factoring or the trivial factoring $n = n \cdot 1$. The trivial factoring occurs when x is increased all the way to $x = \dfrac{n+1}{2}$. For this x value, the corresponding y value will be $y = \dfrac{n-1}{2}$. This can be seen in the following calculations:

$$x^2 - n = \left(\frac{n+1}{2}\right)^2 - n$$

$$x^2 - n = \frac{(n+1)^2}{4} - n$$

$$x^2 - n = \frac{n^2 + 2n + 1}{4} - \frac{4n}{4}$$

$$x^2 - n = \frac{n^2 - 2n + 1}{4}$$

$$x^2 - n = \frac{(n-1)^2}{4}$$

$$x^2 - n = \left(\frac{n-1}{2}\right)^2$$

Thus, $x^2 - n = y^2$. This choice of $x = \dfrac{n+1}{2}$ and $y = \dfrac{n-1}{2}$ yields the trivial factoring, showing that the algorithm will always end:

$$n = (x + y)(x - y)$$

$$n = \left(\frac{n+1}{2} + \frac{n-1}{2} \right) \cdot \left(\frac{n+1}{2} - \frac{n-1}{2} \right)$$

$$n = \left(\frac{2n}{2} \right) \cdot \left(\frac{2}{2} \right)$$

$$n = n \cdot 1$$

Fermat's method of factoring is a nice supplement to the more widely known square root test, and it can even be more efficient than the square root test, as in the cases when the integer n in question only has factors that are close to $\sqrt{n}$. This was shown with our example, $n = 6{,}499$. Furthermore, this simple concrete algorithm provides a nice application of the commonly taught difference-of-two-squares formula.

COMPARING MEANS

It is standard for students to work with the arithmetic sequence of numbers, which is one where there is a common difference between consecutive terms; and with a geometric sequence, which has a common ratio between consecutive terms. An example of an arithmetic sequence would be 1, 5, 9, 13, 17, . . . , where the common difference is 4. An example of a geometric sequence is 1, 5, 25, 125, 625, . . . , where the common ratio is 5. Each of these sequences with a specific endpoint has a midpoint, or average. In the first case, the arithmetic sequence, using only the part 1, 5, 9, 13, 17, the average is the middle number, which is obtained by adding the numbers and dividing by the number of numbers being added: $1 + 5 + 9 + 13 + 17 = \dfrac{45}{5} = 9$. In the case of the geometric sequence   1,   5,   25,   125,   625,   the   geometric   mean   is $\sqrt[5]{1 \cdot 5 \cdot 25 \cdot 125 \cdot 625} = \sqrt[5]{9,765,625} = 25$, which is the nth root of the product of the n numbers.

However, what is neglected as part of the curriculum is the harmonic sequence. This is a very simply constructed sequence, since it merely requires taking the reciprocal of each member of an arithmetic sequence. So, for example, if we consider the earlier mentioned arithmetic sequence and take the reciprocals of each term, we will have a harmonic sequence. One reason why this sequence of numbers is called a "harmonic" is that when you have a set of strings whose lengths form a harmonic sequence and whose tensions are the same, you will get a harmonic sound when you strum them together.

To get the harmonic mean we simply determine the arithmetic mean of the reciprocals of the sequence and then take its reciprocal. In the case above, the harmonic mean of the harmonic sequence $1, \frac{1}{5}, \frac{1}{9}, \frac{1}{13}, \frac{1}{17}$ would be $\dfrac{1}{\frac{1+5+9+13+17}{5}} = \dfrac{1}{9}$.

You might then ask, how would a harmonic mean be of use to us in everyday life? Here's a real-life example. If you buy a number of items at different prices, and want to determine the average price per item, then you simply take the arithmetic mean, or what is typically called the "average." However, if you want to find the average speed of a car traveling to work at a speed of 50 mph and returning over the same route at 30 mph, then taking the arithmetic mean would not be correct since the time spent traveling at 30 mph is considerably greater than the time spent traveling 50 mph over the same route. To find the average speed, then, you would have to use the harmonic mean. In this case, the harmonic mean is

$$\dfrac{1}{\frac{\frac{1}{30}+\frac{1}{50}}{2}} = \dfrac{1}{\frac{30+50}{1500} \cdot \frac{1}{2}} = \dfrac{1}{\frac{4}{150}} = 37\frac{1}{2}.$$

Note that the harmonic mean can be used to find the average of rates as long as they are measured over the same base, such as speed over the same distances or rates of purchase over the same quantity.

You might then ask how these three means compare in size. In order to do this, we shall show how they compare using simple algebra. We will use two numbers, a and b, and find the three means and then compare their magnitudes.

For the two non-negative numbers a and b:

$$(a - b)^2 \geq 0$$
$$a^2 - 2ab + b^2 \geq 0$$

Add $4ab$ to both sides:

$$a^2 + 2ab + b^2 \geq 4ab$$

Take the positive square root of both sides:

$$a + b \geq 2\sqrt{ab}$$
$$\text{or} \quad \frac{a+b}{2} \geq \sqrt{ab}$$

This implies that the *arithmetic mean* of the two numbers a and b is greater than or equal to the *geometric mean*. (Equality is true only if $a = b$.)

Beginning, as we did above, and then continuing along as shown below, we get our next desired comparison—that of the geometric mean and the harmonic mean.

For the two non-negative numbers a and b,

$$(a - b)^2 \geq 0$$
$$a^2 - 2ab + b^2 \geq 0$$

Add $4ab$ to both sides:

$a^2 + 2ab + b^2 \geq 4ab$

$(a + b)^2 \geq 4ab$

Multiply both sides by ab:

$ab\,(a + b)^2 \geq 4a^2b^2$

Divide both sides by $(a + b)^2$:

$ab \geq \dfrac{4a^2b^2}{(a+b)^2}$

Take the positive square root of both sides:

$\sqrt{ab} \geq \dfrac{2ab}{a+b}$. (Remember that $\dfrac{2ab}{a+b}$ is the harmonic mean,

since $\dfrac{2ab}{a+b} = \dfrac{2}{\dfrac{1}{a}+\dfrac{1}{b}}$.)

This tells us that the geometric mean of the two numbers a and b is greater than or equal to the harmonic mean. (Here, equality holds whenever one of these numbers is zero, or if $a = b$). We can, therefore, conclude that arithmetic mean $\geq$ geometric mean $\geq$ harmonic mean.

DIOPHANTINE EQUATIONS

Typically, in an elementary algebra course when an equation is given with two variables, say x and y, it is expected that there will be a second equation with the same two variables so that the two equations can be solved simultaneously. That is, so that what will be sought is a pair of values for these two variables that satisfies both equations. However, what seems to have been omitted from the school curriculum is to show how one can "solve" an equation with two variables when no second equation is pro-vided. Such an equation is often referred to as a *Diophantine equation*, named after the Greek mathematician Diophantus (ca. 201–ca. 285 CE), who wrote about them in a series of books titled *Arithmetica*.

Let's consider such an equation now, which can evolve from a simple problem, such as the following: How many combinations of six-cent and eight-cent stamps can be purchased for five dollars? As we begin, we realize that there are two variables that must be determined, say x and y. Letting x represent the number of eight-cent stamps and y represent the number of six-cent stamps, the equation: $8x + 6y = 500$ should follow. This can then be reduced to $4x + 3y = 250$. At this juncture, we realize that although this equation has an infinite number of solutions, it may or may not have an infinite number of *integral* solutions; moreover, it may or may not have an infinite number of *non-negative integral* solutions (as called for by the original problem). The first problem to consider is whether integral solutions, in fact, exist.

For this a useful theorem may be employed. It states that if the greatest common factor of a and b is also a factor of k, where a, b, and k are integers, then there exist an infinite number of integral solu-tions for x and y in the equation $ax + by = k$. As we mentioned earlier, equations of this type, whose solutions must be integers, are known as Diophantine equations.

Since the greatest common factor of 3 and 4 is 1, which is a factor of 250, there exist an infinite number of integral solutions to the equation $4x + 3y = 250$. The question that we must now consider is, how many (if any) *positive* integral solutions exist for this equation? One possible method of solution is often referred to as *Euler's method* named after the Swiss mathematician Leonhard Euler (1707–1783). To begin, we should solve for the variable with the coefficient of least absolute value; in this case, that is y, which gives us $y = \dfrac{250 - 4x}{3}$. This is to be rewritten to separate the integral parts as

$$y = 83 + \frac{1}{3} - x - \frac{x}{3} = 83 - x + \frac{1-x}{3}.$$

We now introduce another variable, t, and let $t = \dfrac{1-x}{3}$. Solving for x yields $x = 1 - 3t$. Since there is no fractional coefficient in this equation, the process does *not* have to be repeated as it otherwise would have to be (i.e., each time introducing new variables, as with t above). Now substituting for x in the above equation yields $y = \dfrac{250 - 4(1 - 3t)}{3} = 82 + 4t$. For various integral values of t, corresponding values for x and y will be generated. A table of values such as that shown in figure 2.7 might prove useful.

t	...	−2	−1	0	1	2	...
x	...	7	4	1	−2	−5	...
y	...	74	78	82	86	90	...

Figure 2.7

Perhaps by generating a more extensive table, we would notice for what values of t positive integral values for x and y may be obtained. (Remember, for the earlier problem about stamps we seek only positive values for x and y.) However, such a procedure for determining the number of positive integral values of x and y is not very elegant. Therefore, we will solve the following inequalities simultaneously:

$$x = 1 - 3t > 0 \text{ and thus } t < \frac{1}{3}.$$

For the other inequality, $y = 82 + 4t > 0$ and then $t > -20\frac{1}{2}$.

This can then be stated as $-20\frac{1}{2} < t < \frac{1}{3}$.

This indicates that there are twenty-one possible combinations of six-cent and eight-cent stamps that can be purchased for five dollars.

To get a better grasp on this neglected topic, we offer here another Diophantine equation along with its solution to solidify your understanding of this important algebraic procedure. We shall consider solving the Diophantine equation $5x - 8y = 39$.

First, we will solve for x, since its coefficient has the lower absolute value of the two coefficients.

$$x = \frac{y+39}{5} = y+7+\frac{3y+4}{5}$$

Let $t = \frac{3y+4}{5}$, then we solve for y:

$$y = \frac{5t-4}{3} = t-1+\frac{2t-1}{3}$$

Since we still have a fraction in the last equation, let $u = \frac{2t-1}{3}$, and then we solve for t:

$$t = \frac{3u+1}{2} = u + \frac{u+1}{2}.$$

Once again, since there is a fraction in the last equation, we let $v = \frac{u+1}{2}$, and then solve for u: $u = 2v - 1$. We may now reverse the process because the coefficient of v is an integer. Therefore, substituting in reverse order, we get $t = \frac{3u+1}{2}$. Thus, $t = \frac{3(2v+1)+1}{2} = 3v-1$. Furthermore, $y = \frac{5t-4}{3}$. Therefore,

$$y = \frac{5(3v-1)-4}{3} = 5v-3. \text{ Since } x = \frac{y+39}{5}, \text{ we get}$$

$$x = \frac{8(5v-3)+39}{5} = 8v+3.$$

Now, with our values of x and y: $x = 8v + 3$, and $y = 5v - 3$, we have solved the equation and we can then set up a table of values (figure 2.8) as follows:

v	...	−2	−1	0	1	2	...
x	...	−13	−5	3	11	19	...
y	...	−13	−8	−3	2	7	...

Figure 2.8

We notice that since we were not restricted to the positive values of x and y, we have many solutions as shown in the table above. This is an important aspect of algebra dating back to ancient Greek times that also seems to have been left out of the school curriculum.

FALLING SQUARES

At some point your mathematics teachers may have motivated the study of quadratic equations by saying such equations will show up in classical physics. These second-degree equations can be used to describe the motion of objects in the air under the influence of gravity, for example. You may have also been told that the units of measure are important to keep in mind. When we use the units of feet for distance and quarter-seconds for time in considering falling objects, something interesting happens.

Consider dropping a baseball from the top of a very tall building. The object being dropped is unimportant as long as it has a negligible air resistance, hence a baseball is appropriate for our purposes, but a feather would be inappropriate.

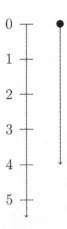

Figure 2.9

If we record the distance s in feet that the baseball has fallen after q quarter-seconds (4 quarter-seconds = 1 second), then we would have the following fact, which we will elaborate on below: $s = q^2$. (See figure 2.9.)

Thus, after 1 quarter-second the baseball will have fallen 1 foot, after 2 quarter-seconds the baseball will have fallen 4 feet, after 3 quarter-seconds the baseball will have fallen 9 feet, and so forth in this pattern of perfect squares.

Why is this true? In physics, the equation for the position s after t seconds due to a constant gravitational acceleration g is: $s = \frac{1}{2}gt^2 + v_0 t + s_0$, where v_0 is the initial velocity of the baseball and s_0 is the initial position. If the ball is simply dropped and not thrown, then the initial velocity is zero, that is, $v_0 = 0$. Counting the starting position as the zero mark from which we measure the distances the baseball has fallen, we get the initial position is zero, written symbolically as $s_0 = 0$. Notice that if we treat the direction toward the ground as the positive direction, then s is measuring the distance the baseball has fallen. The acceleration due to gravity is approximately $g = 32$ feet per second squared. Plugging in these values into the formula, we get

$$s = \frac{1}{2} \cdot 32t^2 + 0 \cdot t + 0$$

$$s = \frac{1}{2} \cdot 32t^2$$

Thus, the equation simplifies as: $s = 16t^2$.

Noting that 1 second is equal to 4 quarter-seconds, we have the equation $q = 4t$ where t is in seconds. In other words, $t = \frac{q}{4}$. Substituting this into the previous expression, we have:

$$s = 16\left(\frac{q}{4}\right)^2, \text{ or}$$

$s = 16 \cdot \dfrac{q^2}{16}$. Simplifying now reveals the reason objects fall with the square pattern noted earlier: $s = q^2$. This elegant pattern that exists in nature is revealed to us through an interesting mixture of knowledge, applying algebra and classical physics. Sometimes it is necessary to broaden your vision in order to appreciate the hidden simplicity in the world around us.

DESCARTES'S RULE OF SIGNS

In all likelihood, you must have encountered an equation of the form $ax^2 + bx + c = 0$, which we know as a quadratic equation. You were able to find the roots by factoring or by using the quadratic formula. But the likelihood of having encountered an equation such as $x^3 - 2x^2 + 3x - 4 = 0$ is clearly less likely. Higher-degree polynomials have roots as well, but unlike the second-degree polynomials, far less is said about them in school. Named after its originator, the French mathematician René Descartes (1596–1650), Descartes's rule of signs will have something to say about general polynomials like this. In particular, we will know something about the positive and negative roots, and all we have to do is count the changes in some plus and minus signs to get this information.

Let's consider an example: $f(x) = x^3 - 2x^2 + 3x - 4$. In this example, we have three sign changes in the coefficients. Reading from left to right, the signs change from the hidden positive coefficient 1 of x^3 to the negative coefficient -2 of x^2, then from -2 to 3, and last from 3 to -4.

Descartes's rule of signs states that the number of positive roots of $f(x)$ either will be equal to the number of sign changes in the coefficients or will be less than the number of sign changes by a multiple of 2. In our example, this means $f(x)$ has either three positive roots, or $3 - 2 = 1$ positive root.

The negative roots of $f(x)$ can be handled in a similar manner.

Instead of working with $f(x)$, we will apply the sign-counting process mentioned earlier to $f(-x)$. Don't let the expression $f(-x)$ scare you. First recall that a negative raised to an even power is positive, while a negative raised to an odd power is negative. Working with $f(-x)$ essentially boils down to flipping the signs of all the terms with odd exponents in $f(x)$. Thus,

$$f(-x) = (-x)^3 - 2(-x)^2 + 3(-x) - 4$$
$$f(-x) = -x^3 - 2x^2 - 3x - 4$$

The signs of the odd exponent terms x^3 and $3x$ in $f(x) = x^3 - 2x^2 + 3x - 4$ were flipped, but the signs of the even power terms stayed the same. Notice there are no sign changes here in $f(-x) = -x^3 - 2x^2 - 3x - 4$. Therefore, Descartes's rule of signs tells us that $f(x) = x^3 - 2x^2 + 3x - 4$ has no negative roots and has either one or three positive roots. This is not an exact answer regarding the nature of the roots like the quadratic formula would offer in the case of second-degree polynomials. In fact, graphing the cubic $y = x^3 - 2x^2 + 3x - 4$ on a graphing calculator or other computing software will reveal just one positive root and no negative roots. (See figure 2.10)

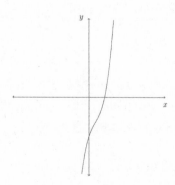

Figure 2.10

We must remember that Descartes lived in the seventeenth century, long before the advent of graphing calculators. In our times, the graphs of such higher-degree polynomials are truly studied in a course in calculus. The simplicity of Descartes's rule of signs allows the curious student a glimpse into the nature of higher-degree polynomials, supplementing what is learned in algebra and precalculus, before he or she eventually engages in the study of calculus. As such, Descartes's rule of signs serves as an interesting bridge spanning these subjects, accessible to any student with the ability to count. Unfortunately, however, this is often omitted from the standard math curricula.

HORNER'S METHOD TO EVALUATE POLYNOMIALS

Polynomials are a familiar topic in algebra courses, where we become exposed to factoring and evaluating polynomials. Horner's method, developed by the British mathematician William George Horner (1786–1837), provides an interesting way to evaluate polynomials at a given value, and this method extends to an alternative way to factor polynomials.

Let's get right to it and consider an example: $f(x) = 2x^3 - x^2 - 7x + 6$. The idea behind Horner's method is to notice that this polynomial can be written in the following nested form:

$$f(x) = [(2x - 1)x - 7]x + 6.$$

Take a moment to expand and simplify the above expression to verify that it is indeed equivalent to the original. If we wish to evaluate $f(x)$ at, say, $x = 3$, then we would essentially be computing:

$$f(3) = [(2 \cdot 3 - 1)\,3 - 7]\,3 + 6.$$

Notice that we will arrive at a compact notation for the following calculations, but first let's see what is happening in the algebraic expressions. Evaluating the innermost parentheses first would yield $2 \cdot 3 - 1 = 5$. Substituting this value 5 into the previous expression would lead to the following: $f(3) = [5 \cdot 3 - 7] 3 + 6$. Now computing $5 \cdot 3 - 7 = 8$ and substituting this value, 8, into the previous expression would lead to the following: $f(3) = 8 \cdot 3 + 6 = 30$.

Thus, we arrive at $f(3) = 30$. The preceding computations can be encoded in the following setup, where we write the value at which the polynomial is to be evaluated at $x = 3$, followed by the coefficients of the original polynomial:

$$3 \, \big| \; 2 \quad -1 \quad -7 \quad 6$$

In this notation, space is left for a second row of numbers above the horizontal line and space is left for a third row of numbers below the horizontal line. The calculations begin by bringing down the leading coefficient, 2, to the third row as follows:

$$
\begin{array}{r|rrrr}
3 & 2 & -1 & -7 & 6 \\
\hline
 & 2 & & &
\end{array}
$$

Next, take the 2 in the third row and multiply by the 3 on the left to get 6. Write the 6 in the second row underneath the -1 as follows:

$$
\begin{array}{r|rrrr}
3 & 2 & -1 & -7 & 6 \\
 & & 6 & & \\
\hline
 & 2 & & &
\end{array}
$$

The -1 and 6 are now to be added, yielding 5, which is written in the third row. Notice that the procedure so far corresponds to starting off with the $2 \cdot 3 - 1 = 5$ calculation from earlier on.

$$
\begin{array}{r|rrrr}
3 & 2 & -1 & -7 & 6 \\
  &   & 6 &   &   \\
\hline
  & 2 & 5 &   &   \\
\end{array}
$$

Repeat the process with the 5 in the third row multiplied with the 3 on the left, and then add the result with −7 to get the next column:

$$
\begin{array}{r|rrrr}
3 & 2 & -1 & -7 & 6 \\
  &   & 6 & 15 &   \\
\hline
  & 2 & 5 & 8 &   \\
\end{array}
$$

This previous step corresponds to computing $5 \cdot 3 - 7 = 8$ from earlier on. The last column corresponds to computing $8 \cdot 3 + 6 = 30$ and yields the desired answer in the last spot of the third row:

$$
\begin{array}{r|rrrr}
3 & 2 & -1 & -7 & 6 \\
  &   & 6 & 15 & 24 \\
\hline
  & 2 & 5 & 8 & 30 \\
\end{array}
$$

The answer, 30, is distinguished from the rest of the third row with a vertical bar as shown above.

Something interesting happens when the polynomial evaluates to 0 at a particular value of $x = a$. In these cases, we know that the polynomial factors with $(x - a)$ as a factor. As an exercise, you may wish to perform Horner's method to evaluate $f(1) = 0$ as follows:

$$
\begin{array}{r|rrrr}
1 & 2 & -1 & -7 & 6 \\
  &   & 2 & 1 & -6 \\
\hline
  & 2 & 1 & -6 & 0 \\
\end{array}
$$

We know that $(x - 1)$ is a factor of $2x^3 - x^2 - 7x + 6$, since the poly-nomial evaluated to 0 at $x = 1$. As a further consequence, the numbers in the third row also indicate that $2x^2 + x - 6$ is a factor, leading to the following incomplete factoring: $2x^3 - x^2 - 7x + 6 = (x - 1)(2x^2 + x - 6)$. This observation leads to the topic of synthetic division.

Typically, students perform long division to obtain

$$\frac{2x^3 - x^2 - 7x + 6}{x - 1} = 2x^2 + x - 6.$$

Synthetic division is an alternative to long division that utilizes the proce-dure for Horner's method shown above. When dividing by the factor $(x - a)$, you may instead use Horner's method to evaluate at $x = a$. The numbers in the third row give the coefficients of the quotient polynomial, and the last number in the third row is the remainder. As another example, using the previous calculations for evaluating $f(x)$ at $x = 3$, we have the following:

$$\frac{2x^3 - x^2 - 7x + 6}{x - 3} = 2x^2 + 5x + 8 + \frac{30}{x - 3}.$$

Horner's method is a nice supplement to your algebraic tool kit, offering you an alternative way to evaluate polynomials. Far from being a mere curiosity, Horner's method actually provides a foundation for syn-thetic division, an alternative to the traditional technique of long division of polynomials. There is more to synthetic division than what we have shown above, and we have not covered all of the various cases, but hope-fully it is clear that Horner's method plays a significant role here and ought to be taught in traditional high school mathematics classes.

GENERATING PYTHAGOREAN TRIPLES

The Pythagorean theorem is perhaps the most famous theorem in math-ematics: the square of the hypotenuse of a right triangle is equal to the sum of the squares of the other two sides. Although most people will recall the Pythagorean theorem as it relates to geometry (as you will see in chapter 3), it still has a lot to do with the equation $a^2 + b^2 = c^2$, which is the algebraic representation of this famous theorem. However, actual Pythagorean triples, which are the integers (a, b, c) that satisfy this equation, are less familiar. Some might remember $(3, 4, 5)$ as the foremost example, and others might add $(5, 12, 13)$ as a second Pythagorean triple stored in memory. Let's discuss a simple technique to actually generate an infinite number of Pythagorean triples.

Start with any positive integer greater than or equal to 3. This starting number will be a in (a, b, c). There are two cases to consider: when a is odd, and when a is even. We will first illustrate the technique with concrete examples, then prove why it works. The following steps in the technique will be justified later, so first let's follow the calculations.

Suppose $a = 7$, an odd number. Square a and then divide by 2, that is, calculate $7^2 = 49$ and $\frac{49}{2} = 24.5$. In this procedure we will round the answer down to get 24. This number, 24, is the second side of the triangle, and this number plus 1 is the hypotenuse. Thus, set $b = 24$ and $c = 24 + 1 = 25$. The generated Pythagorean triple is $(7, 24, 25)$. It is easy to verify that $7^2 + 24^2 = 25^2$ since $49 + 576 = 625$.

Now suppose $a = 8$, an even number. Divide a by 2 and then square, that is, calculate $\frac{8}{2} = 4$ and $4^2 = 16$. This number, 16, minus 1 will be the second side of the triangle, and this number plus 1 is the hypotenuse. Thus, set $b = 16 - 1 = 15$ and $c = 16 + 1 = 17$. The generated Pythagorean triple is $(8, 15, 17)$. It is easy to verify that $8^2 + 15^2 = 17^2$ since $64 + 225 = 289$.

Why does this technique work? Using some algebra will shed light on this situation. We first recall that an odd integer can be written in the form $2m + 1$, while an even integer can be written in the form $2m$, where m is an integer. For example, $7 = 2 \cdot 3 + 1$ and $8 = 2 \cdot 4$.

Suppose $a \geq 3$ is odd, hence $a = 2m + 1$ for some integer m. Square a to get $(2m + 1)^2 = 4m^2 + 4m + 1$. Divide by 2 to get $\frac{(2m + 1)^2}{2} = \frac{4m^2 + 4m + 1}{2}$, which can be rewritten as $2m^2 + 2m + \frac{1}{2}$. Once again, rounding down means dropping the $\frac{1}{2}$ term here. Thus, we set $b = 2m^2 + 2m$ and $c = 2m^2 + 2m + 1$. The Pythagorean triple (a, b, c) can be verified by calculating the squares:

$$a^2 = (2m + 1)^2 = 4m^2 + 4m + 1$$
$$b^2 = (2m^2 + 2m)^2 = 4m^4 + 8m^3 + 4m^2$$

And

$$a^2 + b^2 = (4m^2 + 4m + 1) + (4m^4 + 8m^3 + 4m^2) = 4m^4 + 8m^3 + 8m^2 + 4m + 1$$
$$c^2 = (2m^2 + 2m + 1)^2 = (2m^2 + 2m + 1)(2m^2 + 2m + 1)$$
$$c^2 = 4m^4 + 4m^3 + 2m^2 + 4m^3 + 4m^2 + 2m + 2m^2 + 2m + 1$$
$$c^2 = 4m^4 + 8m^3 + 8m^2 + 4m + 1$$

It is now clear that $a^2 + b^2 = c^2$ is true for the odd case.

Now suppose $a > 3$ is even, hence $a = 2m$ for some integer m. Divide a by 2 to get m. Square it to get m^2. Then we set $b = m^2 - 1$ and $c = m^2 + 1$. The Pythagorean triple (a, b, c) can be verified by calculating the squares:

$$a^2 = (2m)^2 = 4m^2$$
$$b^2 = (m^2 - 1)^2 = m^4 - 2m^2 + 1$$

And

$$a^2 + b^2 = m^4 + 2m^2 + 1$$
$$c^2 = (m^2 + 1)^2 = m^4 + 2m^2 + 1$$

It is once again clear that $a^2 + b^2 = c^2$ for the even case as well.

This technique allows us to generate a Pythagorean triple (a, b, c) that begins with any given integer a greater than or equal to 3. One thing to keep in mind is that there could be other Pythagorean triples that begin with a besides the one calculated by this technique. In other words, this simple technique does not generate all of the Pythagorean triples. There are more general techniques that will handle all of the Pythagorean triples, as you will see below. The beauty of the technique shown here lies in its sheer simplicity. Knowing how to do some basic arithmetic allows you to say much more than just $a^2 + b^2 = c^2$ when you think about the Pythagorean theorem; you can actually generate an infinite number of Pythagorean triples to complement your knowledge of the theorem itself.

The question then arises, how can we more succinctly generate primitive Pythagorean triples (that is, triples in which the numbers a, b, and c do not have a common factor)? More importantly, how can we obtain all Pythagorean triples? That is, is there a formula for achieving this goal? One such formula, attributed to Euclid, generates values of a, b, and c, where $a^2 + b^2 = c^2$ as follows:

$$a = m^2 - n^2$$
$$b = 2mn$$
$$c = m^2 + n^2$$

We will do the simple algebraic task to show that the sum of $a^2 + b^2$ is actually equal to c^2.

$$a^2 + b^2 = (m^2 - n^2)^2 + (2mn)^2$$
$$a^2 + b^2 = m^4 - 2m^2n^2 + n^4 + 4m^2n^2$$
$$a^2 + b^2 = m^4 + 2m^2n^2 + n^4 = (m^2 + n^2)^2 = c^2$$
Therefore, $a^2 + b^2 = c^2$.

If we apply this formula to some values of m and n in the table shown in figure 2.11, we should notice a pattern as to when the triple will be primitive—that is, when the three members are relatively prime, having no common factor other than 1—and also discover some other possible patterns.

m	n	$a = m^2 - n^2$	$b = 2mn$	$c = m^2 + n^2$	Pythagorean Triple (a, b, c)	Primitive
2	1	3	4	5	(3, 4, 5)	Yes
3	1	8	6	10	(6, 8, 10)	No
3	2	5	12	13	(5, 12, 13)	Yes
4	1	15	8	17	(8, 15, 17)	Yes
4	2	12	16	20	(12, 16, 20)	No
4	3	7	24	25	(7, 24, 25)	Yes
5	1	24	10	26	(10, 24, 26)	No
5	2	21	20	29	(20, 21, 29)	Yes
5	3	16	30	34	(16, 30, 34)	No
5	4	9	40	41	(9, 40, 41)	Yes
6	1	35	12	37	(12, 35, 37)	Yes
6	2	32	24	40	(24, 32, 40)	No
6	3	27	36	45	(27, 36, 45)	No
6	4	20	48	52	(20, 48, 52)	No
6	5	11	60	61	(11, 60, 61)	Yes
7	1	48	14	50	(14, 48, 50)	No
7	2	45	28	53	(28, 45, 53)	Yes
7	3	40	42	58	(40, 42, 58)	No
7	4	33	56	65	(33, 56, 65)	Yes
7	5	24	70	74	(24, 70, 74)	No
7	6	13	84	85	(13, 84, 85)	Yes
8	1	63	16	65	(16, 63, 65)	Yes
8	2	60	32	68	32, 60, 68)	No
8	3	55	48	73	48, 55, 73)	Yes
8	4	48	64	80	(48, 64, 80)	No
8	5	39	80	89	(39, 80, 89)	Yes
8	6	28	96	100	(28, 96, 100	No
8	7	15	112	113	(15, 112, 113)	Yes

Figure 2.11

An inspection of the triples in the table would have us make the following conjectures—which, of course, can be proved. For example, the formula $a = m^2 - n^2$, $b = 2mn$, and $c = m^2 + n^2$ will yield primitive Pythagorean triples only when m and n are relatively prime—that is, when they have no common factor other than 1—and *exactly one* of these must be an even number, with $m > n$.

The study of Pythagorean triples can be considered boundless. We have merely scratched the surface here, but once again, for those who wish to pursue this topic further, we recommend the book *The Pythagorean Theorem: The Story of Its Power and Beauty*, by Alfred S. Posamentier (Amherst, NY: Prometheus Books, 2010).

THE FROBENIUS PROBLEM

The German mathematician Ferdinand Georg Frobenius (1849–1917) has a mathematical problem named after him. The famous problem asks for the largest amount of money that cannot be obtained using only coins of specified denominations. Let's consider the problem from specific examples. One problem could be to determine whether we can make $0.37 using US coins. Of course, thirty-seven pennies will trivially suffice, so we will disregard pennies. Is it possible to make $0.37 without using pennies, say, by only using nickels, dimes, and quarters? A moment's thought will reveal that the answer is no, since these coins all have values that are multiples of 5 cents. Now imagine a country in which there are only two coins, one with a value of 5 units and the other with a value of 7 units. What amounts of money can you make using just these two types of coins?

Suppose coin A has a value of 5 units and coin B has a value of 7 units. Is it possible to come up with a combination of these coins that's worth 37 units? Pause and try it yourself before moving on. In algebraic terms, we are looking to find non-negative integers x and y such that $5x + 7y = 37$.

The x and y cannot be negative because it doesn't make sense to count negative amounts of coins. The solution $x = 6$, $y = 1$ works. However, there are some monetary values for which no such combination of coins A and B can be used to represent that value. For example, $5x + 7y = 4$ clearly has no such solutions since the two types of coins are both worth more than 4 units. Once you throw in a single coin, you are already above the desired value of 4 units.

It turns out that $5x + 7y = n$ always has a solution once the positive integer n is large enough. Looking at just the cases for which there are no solutions, there will be a largest such n that yields no solutions. This largest value of n with no solutions is called the Frobenius number of 5 and 7, which we will denote as $g(5, 7)$. In this example, $g(5, 7) = 23$. Later on, we will see there is a simple formula that will give us this calculation.

Given positive integers a and b that are relatively prime—that is, that have no common factors greater than 1—the Frobenius problem asks for the Frobenius number of a and b, or $g(a, b)$. More generally, given m relatively prime positive integers $a_1, a_2, \ldots , a_m$, the Frobenius problem asks for $g(a_1, a_2, \ldots , a_m)$. Here the Frobenius number $g(a_1, a_2, \ldots , a_m)$ is the largest n such that $a_1x_1 + a_2x_2 + \ldots + a_mx_m = n$ has no solutions, where all the x's are non-negative integers.

Some trial and error is customary when first studying the Frobenius problem with small numbers such as $a = 5$ and $b = 7$. Remarkably, there is a simple formula for the case with two generators a and b. The Frobenius number is given by $g(a, b) = ab - a - b$. In our earlier example, $g(5, 7) = 5 \cdot 7 - 5 - 7 = 23$.

Even more remarkably, no one has found a corresponding formula for the case with three (or more) generators, a, b, and c. The simple formula for $g(a, b)$ was known to the English mathematician James Joseph Sylvester (1814–1897) in the late nineteenth century. Ferdinand Georg Frobenius was reputed to have presented this problem of finding a formula for $g(a, b, c)$, or higher, as a challenge to his students, hence the name the *Frobenius problem*. To this day, this challenge is still open, and a corresponding formula for $g(a, b, c)$ is still unknown, over a hundred years later!

The Frobenius problem is an interesting problem from number theory that can be stated in simple terms, using only algebra that is understandable with just a modest mathematical background, and yet this problem has challenged mathematicians for over a century. In the face of simplicity, there sometimes lies an unexpected difficulty carefully hidden beneath the surface. This characteristic seems to be shared by some of the most resilient problems that have managed to challenge and entice generations of mathematicians.

Throughout this chapter we have experienced how algebra allows us to inspect mathematical curiosities in such a way that we can explain why they "work ." The school curriculum can certainly benefit from using these illustrative examples to make mathematics come to life, become useful, and be very motivating.

CHAPTER 3

GEOMETRIC CURIOSITIES

There is a long-standing belief among mathematicians that the geometry course in high school is one of the good indicators of how well students will do in mathematics at the higher levels. Perhaps this is because, unlike algebra, which is often taught as a mechanical process, good geometry teaching instills in the learner logical thought, which is the key to studying higher mathematics. However, geometry is a subject matter vastly more expansive than one year of high school study can cover thoroughly. In this chapter we will explore a number of these often omitted topics and concepts from the field of geometry—realize, though, that we are still merely scratching the surface of this expansive field. As you will see, an exposure to these topics, which clearly enriches your understanding and appreciation of geometry, can go a long way to enhance mathematical thinking among our populace. We will begin with a refreshing look at some of the very basics of geometry; this should set the tone for geometric thinking. We will be covering quite a few topics to which you may have been exposed, but we will approach them from vastly different points of view. For example, we will visit perhaps one of the most popular topics from the geometry course, the Pythagorean theorem. Most people remember the relationship as $a^2 + b^2 = c^2$ and not much else about it, but here we hope to fill in some of the gaps (although by no means fully completing all possible understanding of this most ubiquitous relationship). We hope to do more than just whet your appetite; rather, we hope to motivate you to pursue many of the other wonders of this amazing theorem. You will find many surprises in this chapter, yet despite the vast array of topics presented, your background need not exceed a high school geometry course.

PARALLELOGRAMS AND TRIANGLES

One of the most famous formulas in mathematics is $A = bh$. Many will immediately recognize this formula in the context of rectangles. The area, A, of a rectangle can be computed by multiplying the base length, b, with the height, h. Some will recognize that this formula also gives the area, A, of a parallelogram with base length, b, and height, h. There is a simple way to see why the area formula for a parallelogram is the same as the more familiar area formula for a rectangle.

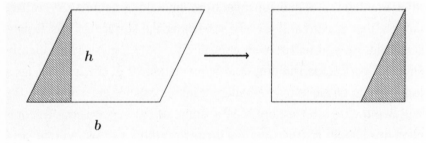

Figure 3.1

The shaded right triangle shown in figure 3.1 can be cut out of the parallelogram and moved to the other side of the parallelogram as shown, forming a rectangle. Notice both the parallelogram and the rectangle are made out of the same shaded and unshaded pieces. The total area of the shaded and unshaded pieces does not change when the pieces are rearranged. Therefore, the area of the parallelogram is the same as the area of the rectangle.

Parallelograms, in turn, can be used to demonstrate the formula for the area of a triangle: $A = \frac{1}{2}bh$.

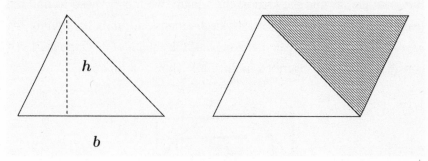

Figure 3.2

Start with a triangle, shown on the left of figure 3.2, with base length *b* and height *h*. Our goal is to form the parallelogram, shown on the right, using two copies of the original triangle, shown on the left. Notice that the shaded copy of the triangle was rotated 180° about the midpoint of the right side of the triangle, while the unshaded copy was not rotated. The shaded and unshaded triangles were joined together on the same corresponding side to form the parallelogram. This common side is the one connecting the top of the original triangle to its lower right vertex.

The parallelogram has area *bh* and the two triangles making up the parallelogram have equal area. Therefore, each of these triangles has an area $A = \frac{1}{2}bh$, as desired.

The humble-looking area formulas for rectangles, parallelograms, and triangles are familiar topics from school. We see that these formulas are also intimately related through the simple demonstrations shown here.

USING A GRID TO CALCULATE AREAS

Typically, the school curriculum presents the topic of finding the area of various geometric shapes through basic area formulas, which are dependent on finding the lengths of the sides of the figures under consideration.

For example, to find the area of the square, we merely need to find the length of one side and square it. An alternative formula for finding the area of the square is to calculate one half the square of its diagonal. Let's consider the square inscribed in the circle shown in figure 3.3.

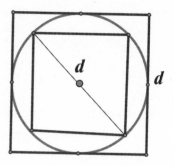

Figure 3.3

The length of the diagonal of the inscribed square is equal to the diameter, d, of the circle. Therefore, the area of the inscribed square is equal to $\frac{1}{2}d^2$, since if we consider each of the four right triangles formed by the two diagonals, each has an area of $\frac{1}{2}\left(\frac{1}{4}d^2\right)$, and the four triangles together then give us the area of the square. The area of the larger circumscribed square, which has a side length d (since it is the same length as the diagonal of the smaller square), has its area as d^2. We can find the area of the circle whose radius is $\frac{1}{2}d$ by simply applying the familiar formula πr^2 for the area of the circle to get $\pi\left(\frac{1}{2}d\right)^2 = \frac{1}{4}\pi d^2$. This will allow us also compare the areas of these three shapes: the inscribed square, the circumscribed square, and the circle. For example, the ratio the area of the inscribed square to that of the circle is

$$\frac{\frac{1}{2}d^2}{\frac{1}{4}\pi d^2} = \frac{2}{\pi}.$$

There is another way to compare these areas, which is by employing a coordinate grid, as shown in figure 3.4.

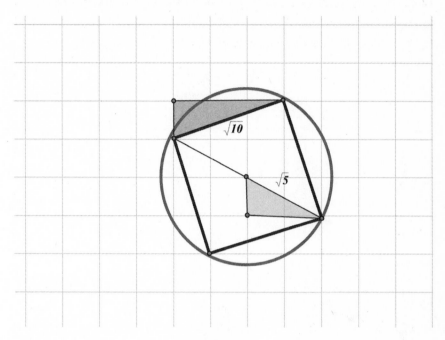

Figure 3.4

Suppose we construct a circle with radius length $\sqrt{5}$ and inscribe a square within the circle. Using the two shaded triangles shown in figure 3.4, we find that the hypotenuse of the upper shaded right triangle—which is also the side length of the inscribed square, length s—can be obtained applying the Pythagorean theorem to the shaded right triangle. We are looking to find the length of the hypotenuse, which is c in the equation. Examining the grid, we can see that $a = 1$ unit on the grid and $b = 3$ units on the grid. Now we apply the Pythagorean theorem. In doing this, we get $1^2 + 3^2 = s^2$, and then $s = \sqrt{10}$. Now we compare the area of the square to that of the circle; we get

$$\frac{\left(\sqrt{10}\right)^2}{\pi\left(\sqrt{5}\right)^2} = \frac{2}{\pi}.$$

There is an alternate procedure for comparing areas, which is usually not presented as part of the standard high school math curriculum. It becomes particularly useful when we wish to compare areas as we do in the following situation. Consider the lines joining vertices and midpoints of the sides of square *ABCD*, as shown in figure 3.5.

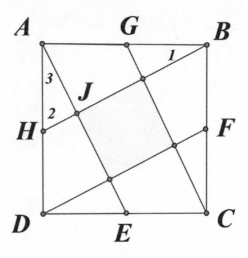

Figure 3.5

Here we are being asked to compare the areas of the smaller (shaded) square to the larger square *ABCD*. One could rightfully ask the question as to why can we assume that the shaded region is, in fact, a square. Clearly the pairs of opposite sides are parallel, which makes it a parallelogram. We can show that $\triangle BAH \cong \triangle ADE$, so that $\angle 1 = \angle 3$. However, $\angle 1$ is complementary to $\angle 2$. Therefore, $\angle 2$ is complementary to $\angle 3$, and consequently $\angle AJH = 90°$; thus, making the smaller shaded quadrilateral a rectangle. However, by the symmetry of the entire figure, we can conclude that this rectangle is in fact a square. Now that we have established that this shaded region is a square, we can come back to the original question about comparing the area of the two squares. Rather than use the method shown above, we will use a grid, as shown in figure 3.6.

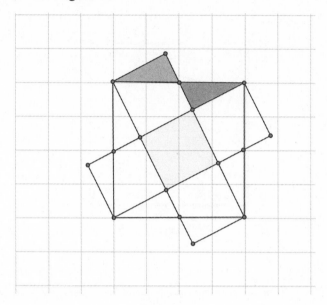

Figure 3.6

Notice that here we have extended each of the segments from the sides of the inside square and through the midpoints of the larger square so that they meet the perpendicular line segments drawn from each of the vertices of the square. We can easily prove that the two shaded triangles at the top of the figure are congruent. With appropriate replacements all the way around each of the four sides of the larger square, we can then show that the oblique cross figure has the same area as the original large square. The rest of the question is rather simple. We can see that the shaded square in the center is $\frac{1}{5}$ of the area of the oblique cross; therefore, it is then also $\frac{1}{5}$ of the area of the larger square.

This technique can be exceedingly useful for somewhat more complex situations. Suppose that instead of using the midpoint of the side of the square, we repeat the same procedure but this time use the trisection points (these are points that partition the line segment into three equal segments), as shown in figure 3.7. Here we are looking for the ratio of the area of the inside square to that of the larger outside square.

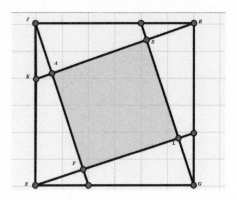

Figure 3.7

Once again, we can see how using a grid can give us the required answer. The following congruencies, shown in figure 3.8, can be easily established: $\triangle AJB \cong \triangle GCB \cong \triangle EDG \cong \triangle EFJ$. This then allows us to easily show that the area of the square *EGBJ* is equal to the area of the figure *ABCDEF*. Then, by counting square units, we can conclude that the area of the inside square (*ASLF*) is $\frac{4}{10} = \frac{2}{5}$ of the area of the figure *ABCDEF*.

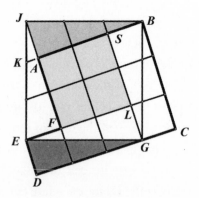

Figure 3.8

This method of comparing areas is typically not shown in the course of the normal school curriculum, yet it can often very much simplify the more traditionally taught methods. Furthermore, it might appeal to learners who have stronger visual-spatial skills than they do number skills.

THE CENTER OF A QUADRILATERAL

When studying geometry, it is well known that the center of a triangle, or balancing point, known as the centroid, is the point of intersection of the medians. Yet locating the center of a quadrilateral seems to have been neglected in our school study of geometry. To close this void in our knowledge of geometry, we shall now consider *two* centers of a quadrilateral. The *centroid* of a quadrilateral is that point on which a quadrilateral of uniform density will balance. This point may be found in the following way. Let M and N be the centroids of $\triangle ABC$ and $\triangle ADC$, respectively. (See figure 3.9.) Let K and L be the centroids of $\triangle ABD$ and $\triangle BCD$, respectively. The point of intersection, G, of MN and KL is the centroid of the quadrilateral $ABCD$.

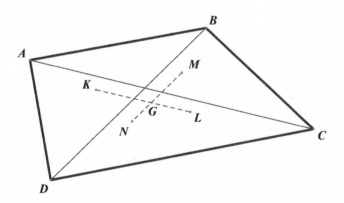

Figure 3.9

In contrast to the centroid, the *centerpoint* of a quadrilateral is the point of intersection of the two segments joining the midpoints of the opposite sides of the quadrilateral. It is the point at which a quadrilateral with equal weights suspended from each of its four vertices would balance. Let's examine a different quadrilateral, *ABCD*. In figure 3.10, *G* is the *centerpoint* of quadrilateral *ABCD*.

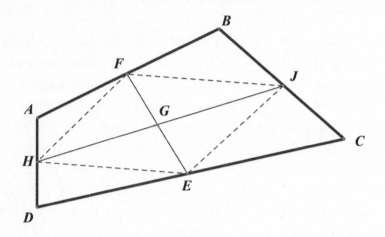

Figure 3.10

We can take this a step further by noting that the segments joining the midpoints of the opposite sides of any quadrilateral bisect each other. This can be justified since these two segments are, in fact, the diagonals of the parallelogram formed by joining the midpoints of the consecutive sides of the quadrilateral, and they bisect each other.

In figure 3.11, points *P*, *Q*, *R*, and *S* are the midpoints of the sides of quadrilateral *ABCD*. We just established that the centerpoint *G* is determined by the intersection of *PR* and *QS*.

An interesting relationship exists between the segments *PR* and *QS* and the segment *MN*, which joins the midpoints *M* and *N* of the diagonals of the quadrilateral. That is, the segment joining the midpoints of the diagonals of a quadrilateral is bisected by the centerpoint.

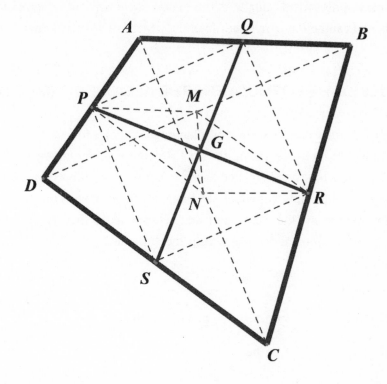

Figure 3.11

This can be justified by considering figure 3.11, where M is the midpoint of segment BD and N is the midpoint of segment AC, and also P, Q, R, and S are the midpoints of the sides of quadrilateral $ABCD$. We then have $\triangle ADC$, with PN as a midline (a line segment joining the midpoints of two sides of a triangle, which is half the length of the third side and parallel to it), which gives us $PN \parallel DC$ and $PN = \frac{1}{2}(DC)$.

Similarly, in $\triangle BDC$, MR is a midline; therefore $MR \parallel DC$ and $MR = \frac{1}{2}(DC)$.

Therefore, $PN \parallel MR$ and $PN = MR$; therefore, it follows that $PMRN$ is a parallelogram.

The diagonals of this parallelogram bisect each other so that MN and PR share a common midpoint, G, which was earlier established as the centerpoint of the quadrilateral. There are many more surprising facts that can be found in quadrilaterals, which we will visit later.

BEYOND THE FORMULA FOR THE AREA OF A TRIANGLE

One of the most remembered topics taught in school is to find the area of a triangle. In its earliest form, the area is determined by taking half the product of its base and height ($Area = \frac{1}{2}bh$). It then is followed by a more sophisticated formula, for which if you are given the lengths of two sides of the triangle and the measure of the included angle, then you can use trigonometry to obtain the area with the formula $Area = \frac{1}{2}ab\sin C$. (See figure 3.12.)

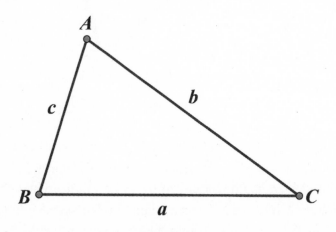

Figure 3.12

Yet too often students are not shown that the area of a triangle can also be determined when all you are given are the lengths of the three sides of the triangle. This void can be filled with Heron's formula, attributed to Heron of Alexandria (10–70 CE), which we encountered earlier in this chapter. Recall that for a triangle whose sides have length a, b, and c,

the area can be found with the formula $Area = \sqrt{s(s-a)(s-b)(s-c)}$,
where the semiperimeter $s = \frac{a+b+c}{2}$.

For example, if we wish to find the area of a triangle whose sides
have lengths 9, 10, and 17, we can use this formula as $s = \frac{9+10+17}{2} = 18$,
and $Area = \sqrt{18 \cdot (18-9)(18-10)(18-17)} = 36$.

Now here is the part that your teacher probably didn't show you.
We know that every triangle can have a circle circumscribed about it—
that is, a circle that touches all three vertices. Suppose we now take the
inscribed triangle and separate two of its sides at one vertex, yet all the
while keeping the two endpoints on the circle, as we show in figure
3.13.

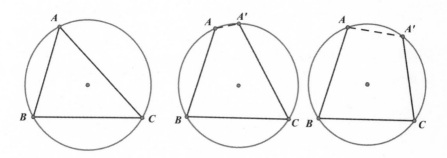

Figure 3.13

That is, we "split" point A into a second point, A'. This allows us to
create a quadrilateral with all four of its vertices on the circle—called
a *cyclic quadrilateral*—and it allows us to extend Heron's formula to
the cyclic quadrilateral by considering the new side, AA', of length d,
so that we now get a formula for the area of a cyclic quadrilateral as
$Area = \sqrt{(s-d)(s-a)(s-b)(s-c)}$. Notice that the $(s-d)$ replaced
the s in Heron's formula for the area of the triangle. This wonderful
formula was first developed by the Indian mathematician Brahmagupta
(598–670 CE). (By the way, he is also believed to have been the first to
compute with zero.)

Therefore, we can now find the area of any cyclic quadrilateral given only the length of its sides. For a cyclic quadrilateral whose sides have lengths 7, 15, 20, and 24, we can use Brahmagupta's formula to get $s = \dfrac{7+15+20+24}{2} = 33$, and then

$$Area = \sqrt{(33-7)(33-15)(33-20)(33-24)} = 234.$$

We should also note that given the four sides of any quadrilateral, it is not possible to determine its area, since such a figure is not stable. That is, if you have four rods of various lengths and form a quadrilateral, there are infinitely many shapes of quadrilateral that can be made. However, with all of its vertices lying on the same circle, the shape is fixed; therefore, the area can be calculated.

Another way to have a quadrilateral with four given side lengths in a unique fixed shape is to be given the measures of one pair of its opposite angles. In this case, the following formula determines the area of the quadrilateral:

$$Area = \sqrt{(s-a)(s-b)(s-c)(s-d) - abcd \cdot \cos^2\left(\frac{A+C}{2}\right)},$$

where A and C are the measures of the opposite angles. Don't be intimidated by this cumbersome formula, but just notice that when the angles at A and C have a sum of 180°, we are left with Brahmagupta's formula, since the quadrilateral with a pair of opposites side supplementary is a cyclic quadrilateral, and in the above formula $\cos^2\left(\frac{180}{2}\right) = 0$.

We may have taken this a bit further than you expected, but this just goes to show you there are lots of connected mathematics concepts, such as these, that would well enrich the standard American high school curriculum.

HERONIAN TRIANGLES

In some classes, teachers may have exposed students to the famous Heron's formula for obtaining the area of a triangle, whose sides have lengths a, b, and c as $Area = \sqrt{s(s-a)(s-b)(s-c)}$, where the semiperimeter $s = \dfrac{a+b+c}{2}$. Clearly, for many triangles—depending on the side lengths— the area could well be an irrational number. The question that could be asked is, For which combination of side lengths will the area of the triangle be an integer number? Such triangles are called *Heronian triangles*.

It will be no surprise that if a right triangle has integer sides, then the area, which is merely one half the product of its legs, will also be an integer number, if at least one of the legs is an even number length. We can create a Heronian triangle by placing two right triangles together that share a common length leg, as shown in figure 3.14.

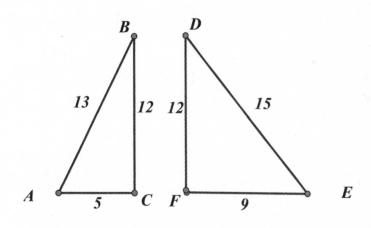

Figure 3.14

Since the area of each of the right triangles is an integer number, namely, 30 and 54, respectively, the area of the entire triangle whose sides are 13, 14, and 15 (that is, triangle formed when sides *BC* and *DF* are juxtaposed) will have an area of 30 + 54 = 84. This can be confirmed by applying Heron's formula: $Area = \sqrt{(21)(21-13)(21-14)(21-15)} = 84$.

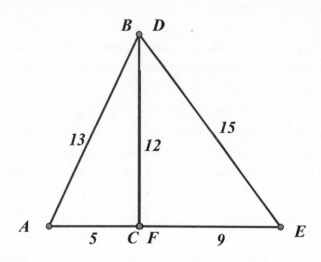

Figure 3.15

Not all Heronian triangles are established by placing two right triangles side by side as we did above. For example, a triangle with side lengths of 5, 29, and 30 has an area of 72, as we can see by applying Heron's formula: $Area = \sqrt{(32)(32-5)(32-29)(32-30)} = 72$, and since none of its altitudes is an integer, placing two right triangles of integer sides together to form this triangle is not possible. If, on the other hand, we consider the right triangles with rational side lengths, then it would be possible to place two right triangles together as we have done above; however, we should note that every altitude of a Heronian triangle must be a rational number length. If we were to split this Heronian triangle into two right triangles, its sides would be of lengths: $\frac{7}{5}, \frac{24}{5}$, 5, and $\frac{143}{5}, \frac{24}{5}$, 29, respectively.

The following is a list of Heronian triangles of integer sides (figure 3.15).

Area	Side Length	Side Length	Side Length	Perimeter
6	5	4	3	12
12	6	5	5	16
12	8	5	5	18
24	15	13	4	32
30	13	12	5	30
36	17	10	9	36
36	26	25	3	54
42	20	15	7	42
60	13	13	10	36
60	17	15	8	40
60	24	13	13	50
60	29	25	6	60
66	20	13	11	44
72	30	29	5	64
84	15	14	13	42
84	21	17	10	48
84	25	24	7	56
84	35	29	8	72
90	25	17	12	54
90	53	51	4	108
114	37	20	19	76
120	17	17	16	50
120	30	17	17	64
120	39	25	16	80
126	21	20	13	54
126	41	28	15	84
126	52	51	5	108
132	30	25	11	66
156	37	26	15	78
156	51	40	13	104
168	25	25	14	64
168	39	35	10	84
168	48	25	25	98
180	37	30	13	80
180	41	40	9	90
198	65	55	12	132

204	26	25	17	68
210	29	21	20	70
210	28	25	17	70
210	39	28	17	84
210	37	35	12	84
210	68	65	7	140
210	149	148	3	300
216	80	73	9	162
234	52	41	15	108
240	40	37	13	90
252	35	34	15	84
252	45	40	13	98
252	70	65	9	144
264	44	37	15	96
264	65	34	33	132
270	52	29	27	108
288	80	65	17	162
300	74	51	25	150
300	123	122	5	250
306	51	37	20	108
330	44	39	17	100
330	52	33	25	110
330	61	60	11	132
330	109	100	11	220
336	41	40	17	98
336	53	35	24	112
336	61	52	15	128
336	195	193	4	392
360	36	29	25	90
360	41	41	18	100
360	80	41	41	162
390	75	68	13	156
396	87	55	34	176
396	97	90	11	198
396	120	109	13	242

You will notice that some of these Heronian triangles have an area that is numerically equal to its perimeter. This is just an extra peculiarity to appreciate. These are the little things that when presented in a classroom would bring the subject to life and make students motivated to search further.

A NEW FORMULA FOR THE AREA OF ISOSCELES TRIANGLES

Elementary geometry was studied extensively in ancient times. In fact, as we mentioned earlier, the famous treatise *Stoicheia* (Greek: *Στοιχεῖα*), written by Euclid of Alexandria (ca. 300 BCE) and today known as Euclid's *Elements* already contained a great deal of what there is to know about geometry. Triangles are the most basic shapes in plane geometry. They can be considered as building blocks for more complex shapes. Every polygon can be decomposed into triangles. Much of what we know about triangles today and probably all of what we learn in school about triangles was known by mathematicians in ancient Greece. Moreover, studying triangles in the plane is not too challenging in the sense that it does not require concepts from higher mathematics. Everyone can take a paper and a pencil, play around with triangles, and figure out formulas for their area, find relationships between certain angles and sides, realize how to construct an inscribed circle or a circumscribed circle, and so on.

Even developing an original proof for the famous theorem of Pythagoras is a feasible task, if you know what the theorem says, and thus, what your task is. Since triangles are such basic shapes that have been studied for thousands of years by innumerable mathematicians, both amateurs and professionals, one would not expect any more new results about triangles. Surprisingly, new discoveries in very old fields of mathematics like the geometry of triangles are still being made, although it happens rarely. From time to time new results are published.

What makes them so appealing is that they had not been recognized (or at least not published) earlier, although they are often so elementary that they could have been found by anyone who went to high school and has fun solving geometry problems.

One such new result was found by the American mathematician Larry Hoehn in 2000. His three-page paper "A Neglected Pythagorean-like Formula" was published in volume 84 of *The Mathematical Gazette*. In the preamble he writes, "This formula has surely been discovered many times, but yet it doesn't seem to appear in the mathematical literature."[1] What is this previously unpublished formula about? Consider an isosceles triangle *ABD* as shown in figure 3.16, where a line segment *BC* perpendicular to *AD* divides the isosceles triangle *ABD* into two right triangles. Then we know by the Pythagorean theorem that $c^2 = a^2 + b^2$.

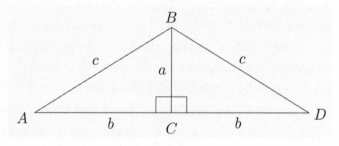

Figure 3.16

Now suppose we move point *C* to the left as it is shown in figure 3.17, where we denote the length of the segment *CD* by *d*. Hoehn claimed that $c^2 = a^2 + bd$, which is a relationship somewhat similar to the Pythagorean theorem. More precisely, it can be interpreted as a generalization of the Pythagorean theorem, since it includes the latter as a special case. Clearly, if the segment *BC* happens to be perpendicular to the segment *AD*, then we have $b = d$ and Hoehn's formula becomes $c^2 = a^2 + b^2$.

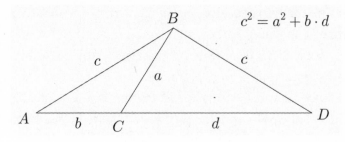

Figure 3.17

To prove the new formula, we first draw the axis of symmetry BE and reflect segment BC about this symmetry axis. (See figure 3.18.)

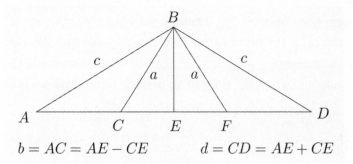

$$b = AC = AE - CE \qquad d = CD = AE + CE$$

Figure 3.18

Then we apply the Pythagorean theorem to the right triangle ABE and obtain $c^2 = AE^2 + BE^2$. From the right triangle BCE we get $a^2 = BE^2 + CE^2$. If we then subtract these two equations, we have

$$c^2 - a^2 = AE^2 + BE^2 - (BE^2 + CE^2) = AE^2 - CE^2 =$$
$$(AE - CE)(AE + CE) = AC \cdot CD = bd.$$

This formula also reveals an interesting relationship between the sides and the diagonals of isosceles trapezoids. To see this, we cut the isosceles triangle shown in figure 3.17 along segment BC and obtain two triangles, as shown in figure 3.19.

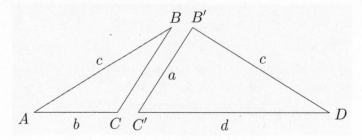

Figure 3.19

Now we reflect triangle ABC about its side AB and attach it to tri-angle $B'DC'$ such that A comes to lie on B' and B comes to lie on D. We obtain an isosceles trapezoid $C'B'CD$ with parallel sides b and d and diagonal c (see figure 3.20). Thus the formula $c^2 = a^2 + bd$ can also be interpreted as a relationship between the diagonal and the sides of an isosceles trapezoid. For example, it can be used to compute the length of the diagonal when the sides of an isosceles trapezoid are given.

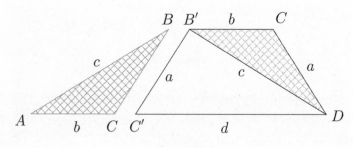

Figure 3.20

Of course, this formula for the length of the diagonal in an isosceles trapezoid can also be derived in a different way. But we wanted to indi-cate that if you discover a relationship for a certain geometrical figure, or, more generally, the solution to a certain mathematical problem, then it is always worth playing around with it and trying to put it in another context. Thereby, it might acquire a whole new meaning or you might be able to exploit it in circumstances that are seemingly completely different from the original problem.

To conclude, elementary geometry still offers a huge playground of mathematical problems waiting to be addressed, and unexpected relationships to be discovered. Everyone can join in the game!

PICK'S THEOREM

The areas of rectangles and triangles are familiar, but what is the area of something more exotic-looking, like the shaded polygon shown in figure 3.21?

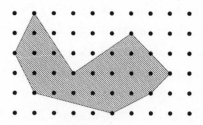

Figure 3.21

When studying geometry and area, it is customary to first understand the areas of basic shapes such as triangles, rectangles, and circles. For the areas of more complicated shapes like the one shown in figure 3.21, a standard practice is to cut up the figure into more manageable smaller basic pieces, then add the areas of the smaller pieces together to obtain the area of the whole.

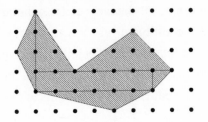

Figure 3.22

Figure 3.22 illustrates this simplifying procedure to reduce the problem into one of counting the areas of triangles and rectangles. It is an exercise to calculate the areas of the five triangles and one rectangle shown, then add it all up to get a total area of 19.5 square units. This process will certainly work in this case to give us the shaded area, but there is another much simpler method available, and it is called *Pick's theorem.*

The dots in the diagrams above are *lattice points*, that is, points in the plane whose x- and y-coordinates are both integers. The x- and y-axes are not important here, so they are omitted in the diagram. A *lattice polygon* is a polygon whose vertices are lattice points. The shaded polygon in figure 3.21 is an example of a lattice polygon.

Pick's theorem gives a simple formula to compute the area of a lattice polygon by counting points in the polygon. The *boundary points*, as the name suggests, are the lattice points on the boundary of the lattice polygon. The boundary points are circled in figure 3.23. Let's define B to be the number of boundary points. In this case, $B = 9$.

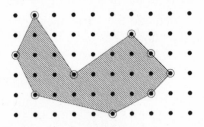

Figure 3.23

The *interior points* are the lattice points contained inside the lattice polygon but not on the boundary itself. The interior points are circled in figure 3.24. Let's define *I* to be the number of interior points. In this case, $I = 16$.

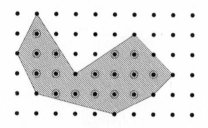

Figure 3.24

Pick's theorem states that the area, *A*, of a lattice polygon can be computed as follows:

$$A = \frac{B}{2} + I - 1$$

In our example, above, the area is: $A = \frac{9}{2} + 16 - 1 = 19.5$.

When faced with a region that's more complicated than the basic shapes we are accustomed to, it is good practice to simplify the situation by decomposing the region into basic shapes. This strategy of reducing the complex to the more familiar and simple permeates much of mathematics. Pick's theorem carries this philosophy even further than one might suspect is possible, reducing the problem of calculating the areas of lattice polygons into one of merely counting dots.

WHEN INTERSECTING LINES MEET A CIRCLE

You may recall the intercept theorem about ratios between various line segments that are created when two intersecting lines are intercepted by a pair of parallel lines. Figure 3.25 shows two lines intersecting at point *P* and their intersection points with a pair of parallel lines, *AB* and *CD*.

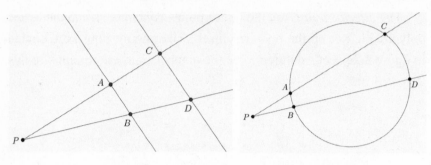

Figure 3.25   Figure 3.26

Stemming from the similarity of triangles APB and CPD, we recall that the statement of the intercept theorem is that $\dfrac{PA}{PC} = \dfrac{PB}{PD}$, that is, the ratio of two segments on the first line through P is equal to the corresponding ratio of two segments on the other line through P. We can also write this relationship as $PA \cdot PD = PB \cdot PC$.

The intercept theorem is concerned with straight lines, the most basic objects of elementary plane geometry. However, there are also other "lines" that are drawn in elementary plane geometry, most importantly circular arcs or complete circles. Have you ever thought about replacing the pair of parallel lines in the intercept theorem with a circle? Then the situation will look like the one depicted in figure 3.26. Is there still any mathematical relationship between PA, PC, PB, and PD? Yes, there is! We have $\dfrac{PA}{PB} = \dfrac{PD}{PC}$, and the cross product of this proportion gives us: $PA \cdot PC = PB \cdot PD$.

This is true for any point P outside or inside the circle and any two lines through P intersecting the circle. If a line, say PAC, is tangent to the circle, then $PA = PC$ (since A and C coincide), but the statement remains true. Before providing a proof of this little theorem, let us formulate it in a slightly different way to emphasize its geometrical meaning:

We are given a circle and a point P outside or inside this circle. If we draw a straight line through P intersecting the circle in points A and C (which might coincide), then the value of the product $PA \cdot PC$ will be the same for all such lines.

Let us now look at the proof. Here we only consider the case that *P* lies outside the circle and both lines intersect the circle twice (that is, they are secants to the circle). But we encourage you to complete the proof for the other possible cases. (The case in which *P* lies inside the circle is shown in figure 3.27.)

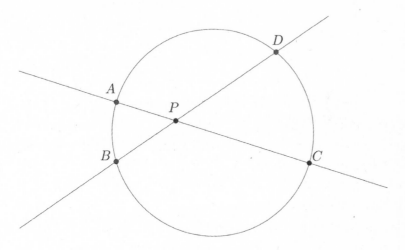

Figure 3.27

First, we draw auxiliary chords *AD* and *BC*, as shown in figure 3.28. Since two angles inscribed in the same arc ($\overset{\frown}{CD}$) are equal, we have $\angle CAD = \angle CBD$. Consequently, their supplementary angles are also equal, namely, $\angle PAD = \angle PBC$. If we now look at the triangles *PAD* and *PBC*, shown in figure 3.29, we can conclude that they are similar triangles. Since corresponding sides of similar triangles are in proportion, we have $\frac{PA}{PB} = \frac{PD}{PC}$, and, therefore, $PA \cdot PC = PB \cdot PD$, which is somewhat different from the situation with the parallel lines but nevertheless comparable.

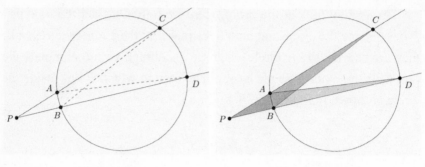

Figure 3.28   Figure 3.29

This exhibits an often neglected relationship between lines and circles, especially in the way it is presented in school courses. We hope that this approach provides you with a deeper insight to this important relationship.

ORIGINS OF TRIGONOMETRY

When trigonometry is introduced at the high school level, the first three functions that are presented are sine, cosine, and tangent. Soon thereafter, an additional three functions are introduced, namely, secant, co-secant and co-tangent. However, the first trigonometric function that we used to set up the initial trigonometric tables was called the chord function. This was initially done by the Greek astronomer Hipparchus (190–120 BCE), who needed to compute the eccentricity of the orbits of the moon and the sun. So he set up a table of values of the chord function. Let's consider this chord function and how it relates to our modern-day trigonometric functions.

In figure 3.30 we have an isosceles triangle with two equal sides of one unit length and the included angle, ϕ. We define the base of this isosceles triangle to have length $chord(\phi)$.

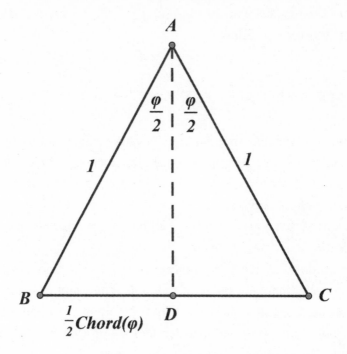

Figure 3.30

If we draw the altitude from vertex A to side BC, we find the length of BD is $\frac{1}{2}chord(\phi)$. Therefore, since $BD = \sin\left(\frac{\phi}{2}\right)$, we have $chord(\phi) = 2\sin\left(\frac{\phi}{2}\right)$. We can also represent the sine function in terms of the chord function as $\sin(\phi) = \frac{1}{2}chord(2\phi)$.

Although Hipparchus's tables produced from this function no longer exist, the oldest surviving tables with trigonometric functions of this sort can be found in Ptolemy's *Almagest*. Working with a high degree of accuracy over two thousand years ago was clearly a remarkable feat. Some of the basic chord functions were $chord(60°) = 1$, and $chord(90°) = \sqrt{2}$. The initial tables ran in increments of $\frac{1}{2}$ degree and were accurate to six decimal places!

These were the beginnings of trigonometry, which over the past few hundred years were supported by separate books containing these detailed tables, albeit for the common trigonometric functions to which we have been exposed at the high school level. Today, the calculator assumes this responsibility.

SINES OF SMALL ANGLES

The study of trigonometry often involves the memorization of the sine and cosine values for certain special angles, such as 30°, 45°, and 60°. For most other angles, a calculator was likely used to find the corresponding trigonometry values. Let's discuss a very nice approximation of sine values that does not require a calculator, but that only works for small angles in radians: $sin\theta \approx \theta$.

We need to first recall the concept of unit circle trigonometry. On the unit circle, that is, a circle of radius length 1 and centered at the origin of the Cartesian plane, start from the point (1,0) and rotate by an angle of θ radians. Rotate counterclockwise for positive angles and clockwise for negative angles. For our purposes here it is important to use radian measure for the angles. The sine value is defined to be the y-coordinate of the point (x,y) on the unit circle you land on after rotating by angle θ, that is, $y = sin\theta$. Similarly, the cosine value is the x-coordinate of this point, written as $x = cos\theta$.

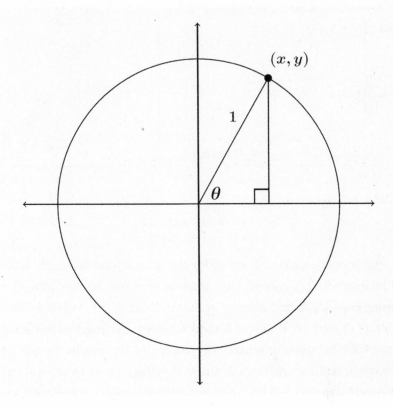

Figure 3.31

The definitions of sine and cosine in unit circle trigonometry evolved from right triangle trigonometry. Notice from the diagram in figure 3.31 that the "opposite side" with respect to the angle θ has length y since it is the height of the triangle. We also notice that the hypotenuse has length 1, since it is the radius of the unit circle. Thus, the sine ratio is "opposite over hypotenuse," which yields $\frac{y}{1} = y$, giving some justification for why $y = sin\theta$ in unit circle trigonometry. Similar reasoning with "adjacent over hypotenuse" applies for $x = cos\theta$.

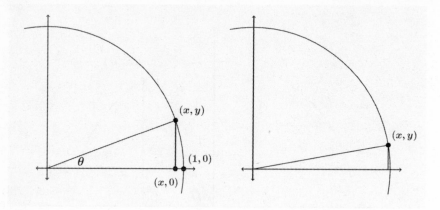

Figure 3.32

Suppose we rotate the radius by just a small positive angle such as in the figure 3.32. Consequently, the point (x,y) will be very close to the starting point $(1,0)$. In this scenario, the small arc of the circle between the points (x,y) and $(1,0)$ is almost vertical. Compare the length of this arc with the vertical line segment between (x,y) and $(x,0)$. The smaller the angle, the more vertical this arc will appear, and so the closer the arc length will match the length of the vertical line segment, y. From earlier, $y = sin\theta$ and so $sin\theta$ is approximately equal to this arc length. The arc length is exactly equal to the small angle θ itself in radian measure because, by definition, the radian measure of the angle θ is given by the arc length divided by the radius of the circle, 1. Thus, we have $sin\theta \approx \theta$ when θ is a small radian-measure angle.

Some slight modifications need to be made for negative θ, but a similar argument shows that the approximation $sin\theta \approx \theta$ works for small negative angles as well.

This simple approximation of sine values for small angles measured in radians is useful in physics, astronomy, and engineering. Furthermore, it provides a wonderful illustration of how radian measure can be useful, a topic whose utility is sometimes not justified in school.

AN UNCONVENTIONAL VIEW OF THE SINE

In school, the sine of an angle is usually introduced as a ratio between the lengths of two sides of a right-angled triangle, namely, the ratio of the side opposite a specified angle to the hypotenuse. However, there also exists a more general geometric interpretation of the sine function that does not require right-angled triangles at all! This different view of the sine function is not often mentioned in school. Not only does it provide a different viewpoint, and, thus, a better understanding of the sine of an angle and its geometrical meaning, but it is also valuable because it provides an effortless proof of the law of sines. Let's now consider this alternative geometric definition of $\sin\alpha$.

In a circle of radius R, let BC be the chord subtended by an angle α at a point on the circle as shown in figure 3.33. Then we will conclude that $\sin\alpha = \dfrac{BC}{2R}$.

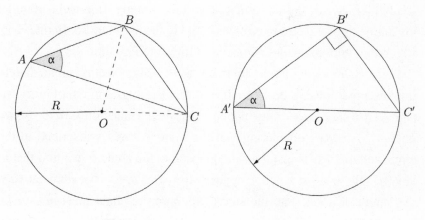

Figure 3.33   Figure 3.34

Before examining the proof, let us take a closer look at the proposed relationship to see if it is plausible that this definition makes sense. Suppose we draw the angle α such that chord $A'C'$ is a diameter of the circle. This situation is depicted in figure 3.34, in which the angle

$\angle A'B'C'$ must then be a right angle so that triangle $A'B'C'$ is a right triangle with hypotenuse $A'C' = 2R$. Thus, for the special situation shown in figure 3.34, our new definition actually coincides with the classical definition of the sine function (that is, the ratio of the length of the opposite side, $B'C'$, to the length of the hypotenuse, $A'C'$. Now for the proof.

We just have to realize that the general case (shown in figure 3.33) can indeed always be reduced to the special situation in which one of the sides of the triangle is a diameter of the circle (as shown in figure 3.34). This is because the inscribed angles $\angle ABC$ and $\angle A'B'C'$ must have the same central angle, that is, $\angle BOC = \angle B'OC'$. This implies that $BC = B'C'$. Therefore, we can now just apply the usual definition of the sine to obtain $\sin\alpha = \dfrac{B'C'}{A'C'} = \dfrac{BC}{2R}$, which is what we wanted to prove.

This relationship reveals a geometric interpretation of the sine of an angle that is valid for arbitrary triangles: For any given triangle, the sine of an angle is always equal to the ratio of the length of its opposite side to the diameter of the circumcircle of this triangle. In a right triangle, the diameter of the circumcircle is equal to the hypotenuse. In this case, we end up with the usual definition of the sine function.

Finally, this alternative approach to the sine function immediately implies a relationship between the sines of the angles in an arbitrary triangle. By a simple rearrangement of terms in the equation $\sin\alpha = \dfrac{BC}{2R}$, we obtain $\dfrac{BC}{\sin\alpha} = 2R$. Considering that the triangle ABC is inscribed in the circle and that BC is just the side opposite to the angle α, we find that in any triangle with sides a, b, c opposite to angles α, β, γ, the ratio between the length of a side and the sine of its opposite angle must always be equal to the diameter of the circumcircle of the triangle. Therefore, $\dfrac{a}{\sin\alpha} = \dfrac{b}{\sin\beta} = \dfrac{c}{\sin\gamma}$, which is called the *law of sines*.

SURPRISING PROOFS OF THE PYTHAGOREAN THEOREM

What do the following three men have in common: Pythagoras, Euclid, and James A. Garfield (1831–1881, the twentieth president of the United States)? All three proved the Pythagorean theorem. The first two should be no surprise, but President Garfield? He wasn't a mathematician. He didn't even study mathematics. As a matter of fact, his only study of geometry, some twenty-five years before he published his proof of the Pythagorean theorem, was informal and was done on his own. In October 1851, he noted in his diary that "I have today commenced the study of geometry alone without class or teacher."[2] While a member of the House of Representatives, Garfield, who enjoyed "playing" with elementary mathematics, developed a clever proof of this famous theorem. It was subsequently published in the *New England Journal of Education* after Garfield was encouraged by two Dartmouth College professors, when he went there to give a lecture on March 7, 1876. The text begins with "In a personal interview with Gen. James A. Garfield, Member of Congress from Ohio, we were shown the following demonstration of the *pons asinorum*, which he had hit upon in some mathematical amusements and discussions with other M.C.'s. We do not remember to have seen it before, and we think it something on which the members of both houses can unite without distinction of party."[3] (See figure 3.35.)

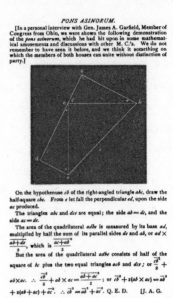

PONS ASINORUM.

[In a personal interview with Gen. James A. Garfield, Member of Congress from Ohio, we were shown the following demonstration of the *pons asinorum*, which he had hit upon in some mathematical amusements and discussions with other M. C.'s. We do not remember to have seen it before, and we think it something on which the members of both houses can unite without distinction of party.]

On the hypotenuse *cb* of the right-angled triangle *abc*, draw the half-square *cbe*. From *e* let fall the perpendicular *ed*, upon the side *ac* produced.

The triangles *abc* and *dce* are equal; the side *ab* = *dc*, and the side *ac* = *de*.

The area of the quadrilateral *adbe* is measured by its base *ad*, multiplied by half the sum of its parallel sides *de* and *ab*, or *ad* × $\frac{ab+de}{2}$, which is $\frac{ac+ab^2}{2}$

But the area of the quadrilateral *adbe* consists of half of the square of *bc* plus the two equal triangles *acb* and *dce*; or $\frac{\overline{cb}^2}{2}$ +

ab×*ac*.   ∴ $\frac{\overline{ab}^2}{2}$ + *ab* × *ac* = $\frac{\overline{ab+ac}^2}{2}$; or $\overline{cb}^2$ + 2(*ab* × *ac*) = $\overline{ab}^2$

+ 2(*ab* + *ac*) + $\overline{ac}^2$.   ∴ $\overline{cb}^2 = \overline{ab}^2 + \overline{ac}^2$. Q. E. D.   [J. A. G.

Figure 3.35

Garfield's proof is actually quite simple, and therefore it can be considered "beautiful." We begin the proof by placing two congruent right triangles ($\triangle ABE \cong \triangle CED$) so that points B, C, and E are collinear as shown in figure 3.36, and that a trapezoid is formed. Notice also that, since $\angle AEB + \angle CED = 90°$, $\angle AED = 90°$, making $\triangle AED$ a right triangle.

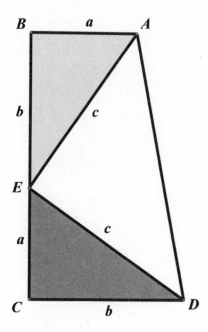

Figure 3.36

Let's examine Garfield's figure, which we have reproduced in figure 3.36. The area of the trapezoid $= \frac{1}{2}$ (sum of bases)·(altitude) $= \frac{1}{2}(a+b)(a+b) = \frac{1}{2}a^2 + ab + \frac{1}{2}b^2$.

The sum of the three triangles (which is also the area of the trapezoid) is $= \frac{1}{2}ab + \frac{1}{2}ab + \frac{1}{2}c^2 = ab + \frac{1}{2}c^2$.

We now equate the two expressions of the area of the trapezoid to get $\frac{1}{2}a^2 + ab + \frac{1}{2}b^2 = ab + \frac{1}{2}c^2$, and then $\frac{1}{2}a^2 + \frac{1}{2}b^2 = \frac{1}{2}c^2$. This then becomes the familiar $a^2 + b^2 = c^2$, which is the Pythagorean theorem.

Of course, it is *possible*, although highly unlikely, that Garfield had knowledge of the tablet shown in figure 3.37 from the Early Han Dynasty (206 BCE–220 CE), which shows the Xuan Tu diagram. The Xuan Tu diagram can produce a proof analogous to Garfield's.

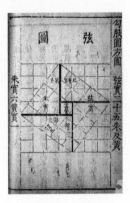

Figure 3.37

The Xuan Tu diagram has also been described as it appears in figure 3.38. If we draw the diagonal of the large center square as shown in figure 3.38 and consider the shaded trapezoid on the right side of the large square, we have the same configuration that guided Garfield.

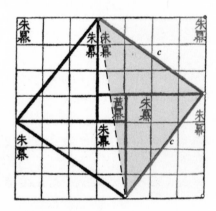

Figure 3.38

Although we credit Pythagoras with having originally developed this foundational theorem of geometry, it is entirely possible that in other cultures this relationship was already known. However, in the Western culture, we still credit Pythagoras. Unfortunately, the school curriculum does not provide enough time to show the many alternative proofs of the Pythagorean theorem. The American mathematician Elisha S. Loomis (1852–1940) published a book in 1940, *The Pythagorean Proposition*,[4] which contains 370 different proofs of the Pythagorean theorem. However, since that time, many more proofs have been published. We provide one such proof that is rarely ever shown in the school curriculum and yet can be easily done, and quite convincingly. Furthermore, historians suppose that Pythagoras used the method of proof that is shown in figures 3.39 and 3.40—perhaps inspired by the pattern of floor tiles.

We begin with a square inscribed inside another square—as shown in figure 3.39. The lengths of the segments are marked as *a*, *b*, and *c*. We notice that the area of the unshaded square is c^2.

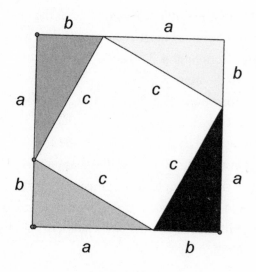

Figure 3.39

We now move the four shaded right triangles and place them within the large square as shown in figure 3.40.

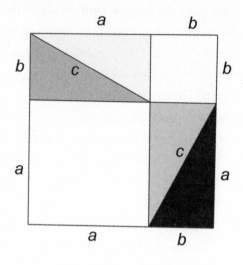

Figure 3.40

When placed in this fashion, we notice that the unshaded areas are two squares, and the sum of those areas is $a^2 + b^2$. By equating the two unshaded regions from figures 3.39 and 3.40, we get $a^2 + b^2 = c^2$, which is the Pythagorean theorem.

Loomis also mentions that the shortest proof of the Pythagorean theorem goes as follows: In figure 3.41 we have a right triangle with an altitude drawn to the hypotenuse. Using the relationship among the three similar right triangles shown in the diagram, we know that each leg of the larger right triangle is the mean proportional between its entire hypotenuse and the nearest segment along the hypotenuse. This allows us to establish the following two equations: Using leg AB of the right triangle, we get

$$\frac{c}{a} = \frac{a}{n}, \text{ or } a^2 = cn,$$

and for leg AC of the right triangle we get

$$\frac{c}{b} = \frac{b}{m}, \text{ or } b^2 = cm.$$

Now by simply adding these two equations, we get

$$a^2 + b^2 = cn + cm = c(n + m) = c^2,$$

which is the Pythagorean theorem as applied to triangle ABC.

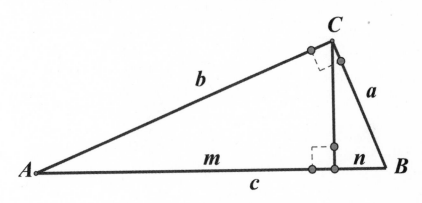

Figure 3.41

There are over four hundred proofs of the Pythagorean theorem available today; many are ingenious, yet some are a bit cumbersome. However, none will ever use trigonometry. Why is this? An astute observer will tell you that there can be no proof of the Pythagorean theorem using trigonometry, since trigonometry depends (or is based) on the Pythagorean theorem. Thus, using trigonometry to prove the very theorem on which it depends would be circular reasoning. We invite you to pursue some of the other clever proofs of the Pythagorean theorem.[5]

BEYOND THE PYTHAGOREAN THEOREM—PART I

You could probably argue that no mathematical relationship has fascinated more people over the centuries than the Pythagorean theorem. As mentioned earlier, there are more than four hundred different proofs available of this famous theorem. Unfortunately, the school curricula do not provide time to explore many of these ingenious proofs. Some of these proofs can be done by simply inspecting a diagram. For example, if we consider the basic idea of the Pythagorean theorem, $a^2 + b^2 = c^2$, we will notice that this can be expressed geometrically by placing a square on each side of a right triangle, as shown in figure 3.42.

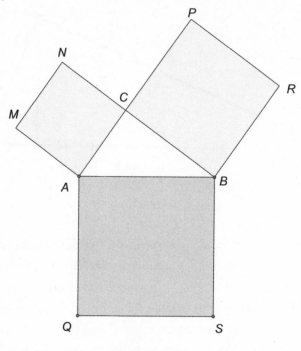

Figure 3.42

From the statement of the Pythagorean theorem, we can conclude that the sum of the areas of the two squares situated on the legs of the right triangle is equal to the area of the square placed on the hypotenuse of the right triangle.

There is a nice demonstration to establish the truth of this statement, which is shown by shifting the areas shown in figure 3.43. This is based on the notion that a parallelogram's area is equal to the product of the base and its height; from which follows that two parallelograms sharing the same base having equal altitudes have the same area (remember a square is also a parallelogram). Follow along progressively from one figure to the next (left to right) and notice how the shaded region gradually shifts from the two squares on the legs of the right triangle to the square on the hypotenuse of the right triangle, thus showing equal areas. We should note that by shifting the shaded regions to the third position, the line *CK*, which is parallel to the two segments *HA* and *GB*, will also be perpendicular to the base of the triangle *ABC*.

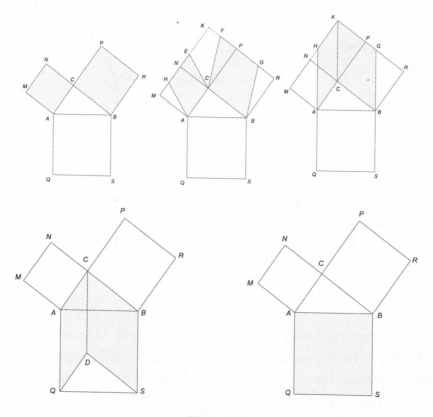

Figure 3.43

If we name the areas of the squares on the sides *a*, *b*, and *c* of the right triangle as S_a, S_b, and S_c, respectively, as shown in figure 3.44, the Pythagorean theorem can be written in the form $S_a + S_b = S_c$.

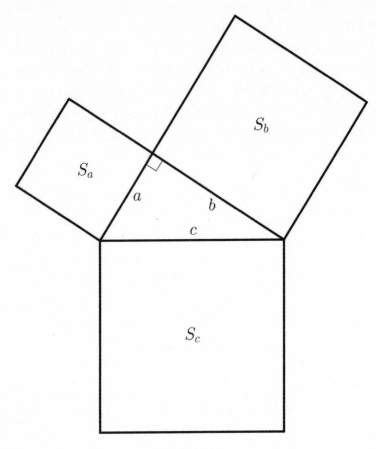

Figure 3.44

Taking a closer look at this leads us quite naturally to a more general result you likely were not introduced to in school. If $S_a + S_b = S_c$, it is quite obvious that $\frac{1}{2} \cdot S_a + \frac{1}{2} \cdot S_b = \frac{1}{2} \cdot S_c$ also holds. Since these expressions are half the areas of the squares on the sides of the triangle, this gives us a relationship between the areas of "half squares." As in the left side of figure 3.45, we can interpret this as a relationship between the isosceles right triangles on the sides.

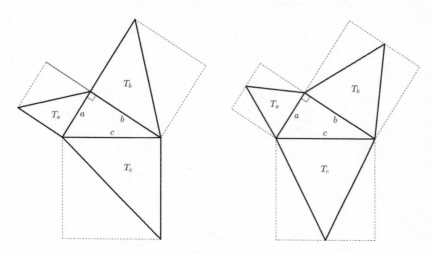

Figure 3.45

Alternatively, as in the right side of figure 3.45, we obtain the same relationship for the isosceles triangle with bases on the sides of the original right triangle. In both cases we have $T_a = \frac{1}{2} \cdot S_a$, $T_b = \frac{1}{2} \cdot S_b$, and $T_c = \frac{1}{2} \cdot S_c$; therefore, $T_a + T_b = T_c$.

In fact, a result of this type is true for any similar figures erected on the sides of a right triangle. The sum of the areas of the two similar figures erected on the legs, a and b, of a right triangle will always equal the area of the similar figure erected on the hypotenuse, c.

The special case of equilateral triangles follows from the situation described in the right side of figure 3.45.

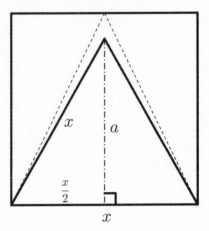

Figure 3.46

Since an equilateral triangle sharing a side-length x with a square (as shown in figure 3.46) has the altitude

$$a = \sqrt{x^2 - \left(\frac{x}{2}\right)^2} = \sqrt{\frac{3x^2}{4}} = \frac{\sqrt{3}}{2} \cdot x,$$

then the area E_x of such a triangle is equal to $E_x = \frac{1}{2} \cdot x \cdot \frac{\sqrt{3}}{2} \cdot x = \frac{\sqrt{3}}{4} \cdot x^2$, and therefore it follows that

$$E_x = \frac{\sqrt{3}}{4} \cdot x^2 = \frac{\sqrt{3}}{2} \cdot \frac{1}{2} \cdot x^2 = \frac{\sqrt{3}}{2} \cdot T_x.$$

From $T_a + T_b = T_c$ in figure 3.45, we therefore obtain

$$\frac{\sqrt{3}}{2} \cdot T_a + \frac{\sqrt{3}}{2} \cdot T_b = \frac{\sqrt{3}}{2} \cdot T_c, \text{ or } E_a + E_b = E_c.$$

Similar arguments hold for any shapes when we erect mutually similar versions of those shapes on the sides of the right triangle. Some examples of this are shown in figure 3.47.

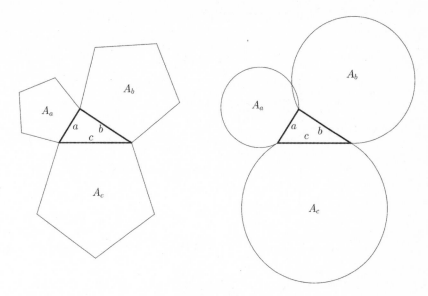

Figure 3.47

In both cases in figure 3.47, the areas have the property $A_a + A_b = A_c$. Of course, there are many other paths we can follow that all start from the Pythagorean theorem. You will encounter a few examples of this in the next two sections.

BEYOND THE PYTHAGOREAN THEOREM—PART II

As we saw in the previous section, the Pythagorean theorem can be extended to consider other similar polygons placed on the legs of a right triangle and on the hypotenuse. The sum of the two areas of the smaller polygons is always equal to the area of the larger polygon. For the particular case when the polygons are equilateral triangles (figure 3.48), we saw in the previous section that the sum of the areas of the two smaller equilateral triangles is equal to the area of the larger one, or $T_3 = T_1 + T_2$.

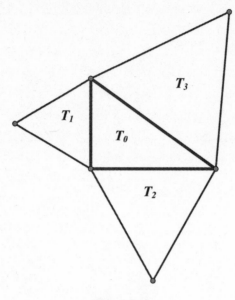

Figure 3.48

We shall now take a giant leap beyond the Pythagorean theorem and change the right angle to an angle of 60°, as shown in figure 3.49.

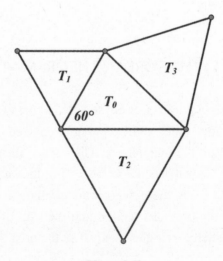

Figure 3.49

When we do this, we end up with a most unusual relationship among the areas of the four triangles: the sum of the areas of the middle triangle and the area of the equilateral triangle, which is on the side opposite the 60° angle, is equal to the sum of the areas of the remaining two equilateral triangles. Symbolically, this can be written as $T_0 + T_3 = T_1 + T_2$. To demonstrate this,[6] we simply rearrange triangles as shown in figure 3.50. We should note that the sum of the three angles that share the vertex on the sides of the large equilateral triangle have a sum of 180°.

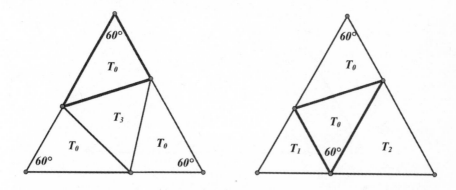

Figure 3.50

The left side of figure 3.50 shows an equilateral triangle composed of three copies of the original center triangle surrounding the triangle that was originally on the side opposite the 60° angle. On the right side of figure 3.50 we have a large equilateral triangle composed of two copies of the original center triangle and the two equilateral triangles that were originally placed on the sides composing the original 60° angle. These two equilateral triangles, on the left side and on the right side of figure 3.50, are equal in area. We know this because each of their side lengths are the sum of the lengths of the two sides that included the 60° angle of the original triangle. This allows us to set up the following area relationship using the triangle designation shown in figure 3.50. That is, $3T_0 + T_3 = 2T_0 + T_1 + T_2$, which reduces to the following lovely unexpected relationship: $T_0 + T_3 = T_1 + T_2$.

Analytically, this is a simple consequence of the law of cosines. If the sides of the center triangle from figure 3.49 are named a, b, and c (with sides a and b enclosing the 60° angle), the law of cosines gives us $c^2 = a^2 + b^2 - 2ab\cos 60° = a^2 + b^2 - 2ab \cdot \dfrac{1}{2} = a^2 + b^2 - ab$, or $ab + c^2 = a^2 + b^2$. Taking a closer look at a triangle with an angle of 60° enclosed by sides of length x and y as in figure 3.51, we recall that the area of the triangle is equal to $T = \dfrac{1}{2} \cdot x \cdot a_x = \dfrac{1}{2} \cdot x \cdot y \cdot \sin 60° = \dfrac{\sqrt{3}}{4} \cdot xy$.

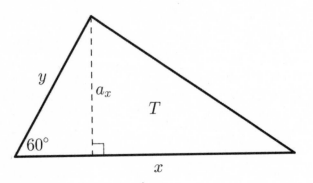

Figure 3.51

All triangles in figure 3.49 are of this type, and we, therefore, have $T_o = \dfrac{\sqrt{3}}{4} \cdot ab$, $T_1 = \dfrac{\sqrt{3}}{4} \cdot a^2$, $T_2 = \dfrac{\sqrt{3}}{4} \cdot b^2$, and $T_3 = \dfrac{\sqrt{3}}{4} \cdot c^2$. But, of course, we derived $ab + c^2 = a^2 + b^2$ from the law of cosines, and multiplying this, indeed, gives us $b + \dfrac{\sqrt{3}}{4} \cdot ab + \dfrac{\sqrt{3}}{4} \cdot c^2 = \dfrac{\sqrt{3}}{4} \cdot a^2 + \dfrac{\sqrt{3}}{4} \cdot b^2$, or $T_o + T_3 = T_1 + T_2$, as previously claimed.

If you are feeling ambitious, you may wish to show that if the 60° angle of the original triangle is changed to 120°, then $T_3 = T_0 + T_1 + T_2$, while if the angle of the original triangle is changed to 30°, then the following relationship evolves: $3T_0 + T_3 = T_1 + T_2$. On the other hand, if that critical angle is enlarged to 150°, then we get $T_3 = 3T_0 + T_1 + T_2$.

This is just one of the almost-boundless extensions that can be attributed to the Pythagorean theorem, but there was likely not enough time in the school curriculum to have it properly presented and explored.

BEYOND THE PYTHAGOREAN THEOREM—PART III

You won't need special glasses to appreciate the step we take in this section, although we are moving from the two-dimensional plane to three dimensions. In fact, we will even peer over the edge into higher dimensions, all with the aid of the Pythagorean theorem. But before we get ahead of ourselves too much, let us take a step back and recall what the Pythagorean theorem tells us about rectangles.

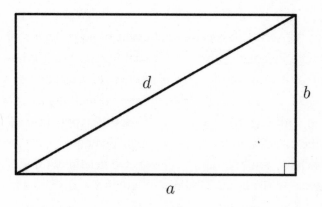

Figure 3.52

As we see in figure 3.52, a diagonal with the length d divides a rectangle with sides of length a and b into two congruent right triangles. The Pythagorean theorem, therefore, tells us that $a^2 + b^2 = d^2$ holds for these segments.

Now, let us go up a dimension. The analogous geometric figure in three dimensions to the rectangle in two dimensions is the cuboid (or rectangular parallelepiped). As is the case for the rectangle in two dimensions, the cuboid's edges (and sides) can only be parallel or perpendicular (although they need not have a common point even if they are not parallel). In figure 3.53 we see such a cuboid with edges of length a, b, and c and a diagonal of length d.

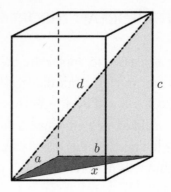

Figure 3.53

Just as the rectangle has two diagonals of equal length, the cuboid has four such diagonals, also all of equal length. We can now express this length d in terms of the lengths of the edges a, b, and c. As we see in figure 3.53, there is a right triangle (represented in dark gray) in the base plane with legs of length a and b and a hypotenuse of length x, which is the diagonal of the base rectangle. Of course, we have $a^2 + b^2 = x^2$. We also have a right triangle (represented in a lighter gray) in a plane perpendicular to the base, with legs of length x and c and a hypotenuse of length d. Therefore, we obtain $d^2 = x^2 + c^2 = a^2 + b^2 + c^2$.

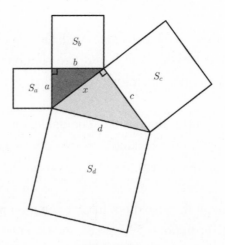

Figure 3.54

We see that the two-dimensional expression $a^2 + b^2 = d^2$ has a very similar counterpart in the "next higher" three-dimensional case, namely $a^2 + b^2 + c^2 = d^2$. As in the two-dimensional case, we can look for a geometric interpretation of this. In figure 3.54, we see the two triangles used in the above calculations, with sides of a, b, x and x, c, d, respectively, rotated into a common plane. If we erect a square on each of the sides of the resulting (shaded) quadrilateral, their areas, S_a, S_b, S_c and S_d, are equal to a^2, b^2, c^2, and d^2, respectively, and we have $S_a + S_b + S_c = S_d$, which again reminds us of the equation $S_a + S_b = S_c$ from Part I.

If we prefer, we can find an equivalent interpretation in terms of volume. We are thinking about a three-dimensional configuration after all, so it is only appropriate that there would be such a thing. If we think of figure 3.54 as the base of a solid configuration, a prism with depth d, we see that the sum of the volumes of the cuboids with base areas S_a, S_b, and S_c is equal to the volume of the cube with base S_d. This results analytically by multiplying $S_a + S_b + S_c = S_d$ by d:

$$S_a + S_b + S_c = S_d \Leftrightarrow a^2 + b^2 + c^2 = d^2 \Leftrightarrow a^2d + b^2d + c^2d = d^3.$$

This step from two to three dimensions can be taken even further. If we can have a shape in two dimensions with edges in two mutually perpendicular directions and a shape in three dimensions with edges in three mutually perpendicular directions, what is to stop us from imagining a "four-dimensional" shape with edges in four mutually perpendicular directions? Of course, our own real world is three-dimensional, but when has mathematical abstraction ever allowed itself to be hindered by physical reality? In order to imagine a "four-dimensional space" in which such an object is possible, we must require the continuation of the Pythagorean theorem in this space. If we have a four-dimensional object with mutually perpendicular edges of length a, b, c, and e and a "diagonal" of length d, we require $a^2 + b^2 + c^2 + e^2 = d^2$ to hold.

Of course, there is more to it than that. If we want to introduce the concept of a four-dimensional space properly, mathematical precision requires a lot more than what we have offered here. It is true, however, that higher-dimensional geometries, analogous to our usual two-dimensional and three-dimensional Euclidean geometries, can be defined in such a way. While this concept of higher dimensions is not quite the same as that of four-dimensional space-time, many of us are at least a little bit familiar with this from general relativity. It is a quite fascinating (and surprisingly useful!) idea to think about the properties of such a four-dimensional space, or even its higher-dimensional cousins. Sadly, this is not part of a school mathematics preparation.

THE PYTHAGOREAN THEOREM
EXTENDED TO THREE DIMENSIONS

In the concluding section regarding the Pythagorean theorem, we will take another look at it from a three-dimensional aspect. Imagine cutting a corner off a rectangular solid. The piece represented by the corner is a tetrahedron with three of its faces as right triangles. In figure 3.55 you will find such a tetrahedron for which point P is the vertex of the original rectangular solid. A lovely extension of the Pythagorean theorem can be shown on this geometric solid. Namely, the sum of the squares of the areas of the three right-triangle faces is equal to the square of the area of the triangle representing the remaining face of the tetrahedron. Using the tetrahedron pictured in figure 3.55, we have the following: $(Area\ X)^2 + (Area\ Y)^2 + (Area\ Z)^2 = (Area\ \triangle ABC)^2$, which is a nice analog to the Pythagorean theorem. This relationship is known as De Gau's theorem, named after the French mathematician Jean Paul de Gau de Malves (1712–1785), although it appears to have been known to the German mathematician Johann Faulhaber (1580–1635) and the famous French mathematician René Descartes (1596–1650).

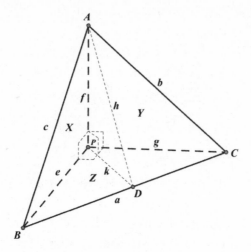

Figure 3.55

To justify this wonderful relationship, we begin by considering the areas of the three right-triangle faces of the tetrahedron, shown in figure 3.55:

$$Area\ X = \frac{ef}{2}, \ Area\ Y = \frac{fg}{2}, \text{ and } Area\ Z = \frac{eg}{2}.$$

We then construct a plane through AP that cuts triangle BPC so that line PD is perpendicular to BC. We know that the area of triangle ABC is $\frac{ah}{2}$, and from the Pythagorean theorem, we have $h^2 = f^2 + k^2$, and the area of triangle BPC is equal to $\frac{ak}{2}$. We can now write the following steps with what we have so far: $\left(Area\ \triangle ABC\right)^2 = \left(\frac{ah}{2}\right)^2 = \frac{a^2h^2}{4}$. It then follows that:

$$4\left(Area\ \Delta ABC\right)^2 = a^2 h^2$$
$$= a^2\left(k^2 + f^2\right)$$
$$= a^2 k^2 + a^2 f^2$$
$$= 4\left(Area\ \Delta BPC\right)^2 + a^2 f^2$$
$$= 4\left(Area\ \Delta BPC\right)^2 + \left(e^2 + g^2\right) f^2$$
$$= 4\left(Area\ \Delta BPC\right)^2 + e^2 f^2 + g^2 f^2$$
$$= 4\left(Area\ \Delta BPC\right)^2 + 4\left(Area\ \Delta BPA\right)^2 + 4\left(Area\ \Delta APC\right)^2$$

Thus, we have established this rather amazing extension of the Pythagorean theorem into the third dimension. Again, see how the Pythagorean theorem, both in two dimensions and beyond, is a key building block within the field of geometry. If you are interested in pursuing this amazing theorem further, we recommend the book *The Pythagorean Theorem: The Story of Its Power and Beauty*, by A. S. Posamentier (Amherst, NY: Prometheus Books, 2010).

POLYHEDRA: SIDES, FACES, AND VERTICES

In the school curricula we seem to focus on the geometry as it applies to lengths of lines and arcs, similarity and congruence, area and volume, and so on. However, what we don't do is focus on the relationship of the sides, the faces, and the vertices of various solid shapes, or specifically polyhedra. The famous Swiss mathematician Leonhard Euler (1707–1783) discovered a lovely relationship among the number of vertices, faces, and edges of polyhedra (which are basically geometric solids). Before delving further into this discussion, you might want to identify a number of solids and count the number of vertices (V), faces (F), and edges (E), and make a chart of these findings. Do you see a pattern emerging? You ought to discover that for all of these figures, the following relationship holds true: $V + F = E + 2$.

Let's consider, for example, a cube, for which there are eight vertices, six faces, and twelve edges. This fits Euler's formula: $8 + 6 = 12 + 2$.

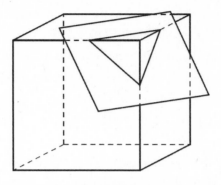

Figure 3.56

When we pass a plane through a corner of the cube, as shown in figure 3.56, we separate one of the vertices from the rest of the polyhedron (in this case the cube) and create a trihedron (the corner section cut from the cube). But, in the process of removing one of the vertices, we added to the polyhedron one face, three edges, and three new vertices. If V is increased by 2 ($-1 + 3 = 2$), F is increased by 1, and E is increased by 3, then $V - E + F$ remains unchanged, and the formula still holds true:

$$V + F = E + 2 = (8 + 2) + (6 + 1) = (12 + 3) + 2.$$

We can obtain a similar result for any polyhedral angle, since there is no change in the expression $V - E + F$.

We know the Euler formula applies to a tetrahedron, which is the "cut off" pyramid, which has $V + F = E + 2$, as shown by $4 + 4 = 6 + 2$. From the above argument, we can conclude that it applies to any polyhedron that can be derived by passing a plane that cuts off a vertex and eliminates a tetrahedron a finite number of times. However, we would like it to apply to all simple polyhedra. In the proof, we would need to show that in regard to the value of the expression $V - E + F$, any polyhedron agrees with the tetrahedron. To do this we need to discuss a relatively new branch of mathematics called topology.

Topology is a very general type of geometry. Establishment of Euler's formula is a topological situation. Two figures are said to be topologically equivalent if one can be made to coincide with the other by distortion—shrinking, stretching, or bending—but not by cutting or tearing. Here is an example of how we can distort one object into another, by molding it differently but without cutting or tearing the whole. Using this method, for instance, we can see that a teacup and a doughnut are topologically equivalent. (The hole in the doughnut becomes the inside of the handle of the teacup, and the rest of the teacup can be molded from the "dough.")

Topology has been called "rubbersheet geometry." If a face of a polyhedron is removed, the remaining figure is topologically equivalent to a region of a plane. We can deform the figure until it stretches flat on a plane. The resulting figure does not have the same shape or size, but its boundaries are preserved. Edges will become sides of polygonal regions. There will be the same number of edges and vertices in the plane figure as in the polyhedron. Each face of the polyhedron, except the one that was removed, will be a polygonal region in the plane. Each polygon, which is not a triangle, can be cut into triangles, or triangular regions, by drawing diagonals. Each time a diagonal is drawn, we increase the number of edges by 1 but we also increase the number of faces by 1. Hence, the value of $V - E + F$ is undisturbed.

Triangles on the outer edge of the region will have either one edge on the boundary of the region, such as triangle ABC in figure 3.57, or two edges on the boundary, such as triangle DEF. We can remove triangles, such as triangle ABC, by removing the one boundary side, for example, AC. This decreases the faces by 1 and the edges by 1, and leaving $V - E + F$ unchanged. If we remove the other kind of boundary triangle, such as triangle DEF, we decrease the number of edges by 2, the number of faces by 1, and the number of vertices by 1. Once again, $V - E + F$ *is* unchanged. This process can be continued until one triangle remains.

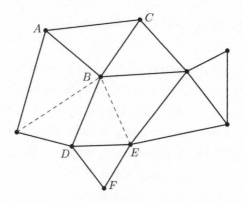

Figure 3.57

The single triangle has three vertices, three edges, and one face. Hence, $V - E + F = 1$. Consequently, $V - E + F = 1$ in the plane figure obtained from the polyhedron by distortion. Since one face had been eliminated, we conclude that for the polyhedron, $V - E + F = 2$.

This procedure applies to any simple polyhedron. An alternative to the approach of distorting the polyhedron to a plane after a face has been eliminated can be named "shrinking a face to a point." If a face is replaced by a point, we lose the n edges of the face and the n vertices of the face, and we lose a face and gain a vertex (the point that replaces the face). This leaves $V - E + F$ unchanged. This process can be continued until only four faces remain. Then any polyhedron has the same value for $V - E + F$, as does a tetrahedron. The tetrahedron has four faces, four vertices, and six edges: $4 - 6 + 4 = 2$. Here is a different type of geometry, and it, too, offers some refreshing delights, perhaps a bit beyond the standard curriculum.

LUNES AND THE RIGHT TRIANGLE

The area of a circle is not typically commensurate with the areas of rectilinear figures. That is, it is quite unusual to be able to construct a circle equal in area to a rectangle, or a parallelogram, or for that matter any other figure composed of straight lines, which we typically refer to as "rectilinear" figures. However, with the help of the Pythagorean theorem we can construct a figure composed of circular arcs that has an area equal to a triangle. You see, the inclusion of π in the circle-area formula usually causes a problem of equating circular areas to noncircular areas, the latter of which do not include π. This is the result of the nature of π, an irrational number that can rarely be compared to rational numbers. Yet we will do just that here.

Let us consider a rather odd-shaped figure, a lune, which is a crescent-shaped figure (such as that in which the moon often appears) formed by two circular arcs. Recall, the Pythagorean theorem states that the sum of the areas of the squares on the legs of a right triangle is equal to the area of the square on the hypotenuse. As a matter of fact, as we mentioned earlier, we can easily show that the "square" can be replaced by any similar figures (drawn appropriately) on the sides of a right triangle: The sum of the areas of the similar shapes on the legs of a right triangle is equal to the area of the similar shape on the hypotenuse, such as the two examples shown in figure 3.58.

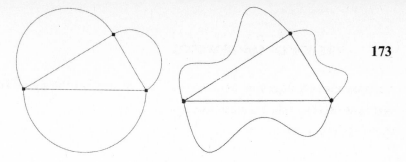

Figure 3.58

This can then be restated for the specific case of semicircles (which are, of course, similar) to read: The sum of the areas of the semicircles on the legs of a right triangle is equal to the area of the semicircle on the hypotenuse. Thus, for figure 3.59, we can say that the area of the semicircles are related as follows:

$$Area \ P = Area \ Q + Area \ R.$$

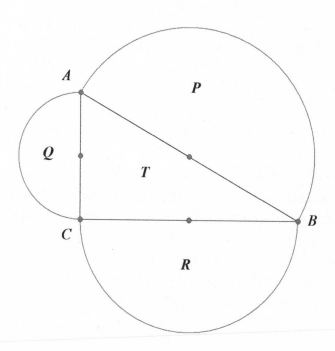

Figure 3.59

Suppose we now flip semicircle P over the rest of the figure (using AB as its axis) as shown in figure 3.60.

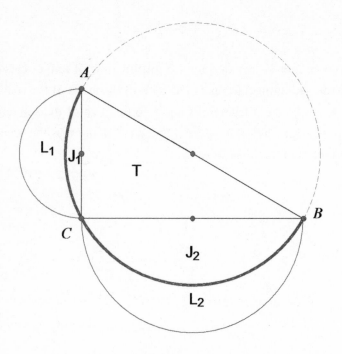

Figure 3.60

Let us now focus on the lunes formed by the two semicircles. We mark the lunes L_1 and L_2, indicated by the unshaded regions in figure 3.61.

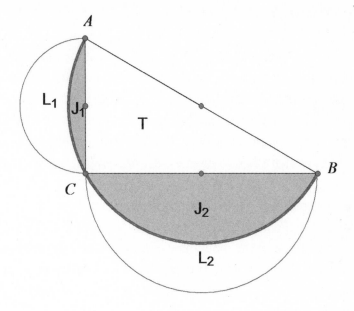

Figure 3.61

Earlier (in figure 3.59) we established that *Area P = Area Q + Area R*. In figure 3.61 that same relationship can be written as follows:

Area J$_1$ + Area J$_2$ + Area T = (Area L$_1$ + Area J$_1$) + (Area L$_2$ + Area J$_2$).

If we subtract *Area J$_1$ + Area J$_2$* from both sides of the equation, we get the astonishing result that *Area T = Area L$_1$ + Area L$_2$.*

That is, we have a rectilinear figure (the triangle) equal to some non-rectilinear figures (the lunes). This is quite unusual since the measures of circular figures seem to always involve π, while rectangular (or straight-line) figures do not, and these are known to be incommensurate! Here again we have another fascinating aspect of fairly basic mathematics that is overlooked or left out of the standard high school math course.

CONCURRENCY

The topic of concurrency in geometry is typically presented in the context of showing that each of the three sets of altitudes, medians, and angle bisectors of a triangle determines a common point of intersection. Of course, in an equilateral triangle, these three points of concurrency converge to one point. However, there are lots of interesting points of concurrency that seem to have evaded the typical school geometry curriculum.

For example, there is the famous theorem attributed to the Italian mathematician Giovanni Ceva (1647–1734). In 1678, Ceva proved a most remarkable theorem about three concurrent lines in a triangle. His theorem states the following: The three lines containing the vertices of triangle ABC (figure 3.62) and intersecting the opposite sides in points L, M, and N, respectively, are concurrent if and only if $\frac{AM}{MC} \cdot \frac{BN}{NA} \cdot \frac{CL}{BL} = 1$.

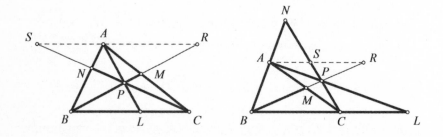

Figure 3.62

The proof of this theorem uses standard geometric concepts.[7] (See the appendix for a proof of Ceva's theorem.)

A simple way to look at Ceva's theorem is to notice that the product of the lengths of the alternating segments along the sides of the triangle are equal. That is, for triangle ABC in figure 3.62, we would state that as $AM \cdot BN \cdot CL = MC \cdot NA \cdot BL$.

Another point of concurrency that is very simple to demonstrate and proved very easily using Ceva's theorem is often referred to as the Gergonne point of a triangle. In the standard geometry course, we learn that the center of an inscribed circle is determined by the point of concurrency of the angle bisectors of the triangle. However, the inscribed circle also helps us determine another point of concurrency relative to the triangle. In figure 3.63, we notice that the lines joining the points of tangency of an inscribed circle of the triangle and the opposite vertices determine a point of concurrency, which we call the Gergonne point, and named after its discoverer, the French mathematician Joseph Diaz Gergonne (1771–1859).

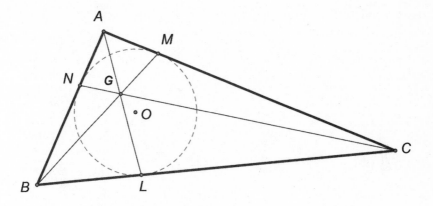

Figure 3.63

Analogously, suppose we now take escribed circles, that is, circles that are tangent to the three sides of the given triangle yet lie outside of the triangle, and join the points of tangency with the opposite vertices of the given triangle (as shown in figure 3.64). Here we see the lines *AD*, *BE*, and *CF* are concurrent at point *X*.

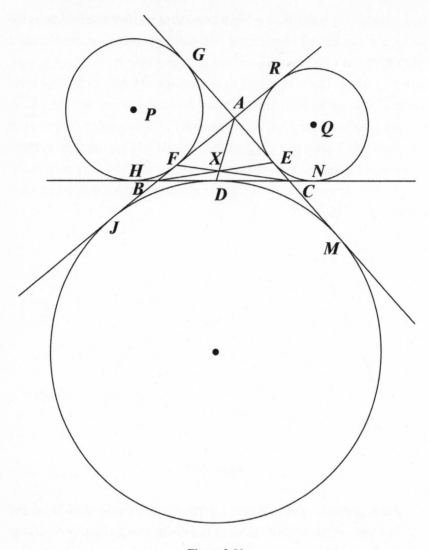

Figure 3.64

The topic of concurrencies can be extended from triangles to circles. For example, consider a theorem developed in 1838 by the French mathematician August Miquel (1816–1851), who found that circles containing a vertex and a common point on each of the sides of the triangle are concurrent at a point. In figure 3.65, we see that circles with centers P, Q, and R each contain a common point on the sides of a triangle and a vertex of the triangle and consequently meet at a common point M, often called the Miquel point.

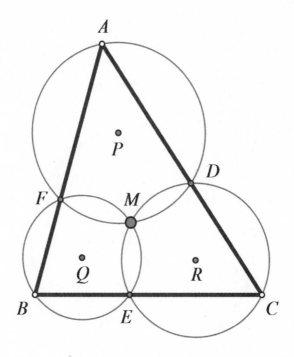

Figure 3.65

Examples of concurrency are numerous and almost boundless, and yet much of their existence seems to have been neglected when it comes to the school curriculum. Perhaps there is just not enough time to enjoy these remarkable aspects of geometry, or perhaps teachers are simply unfamiliar with these gems in geometry. Let's take one more example, since it is quite astonishing.

Consider three equally sized circles with centers C_1, C_2 and C_3 and radius r drawn to meet at a point P. (See figure 3.66.) The remarkable aspect here is that the three points of intersection—points A, B, and

D—determine a circle with center *C* and a radius equal to that of the three original circles.

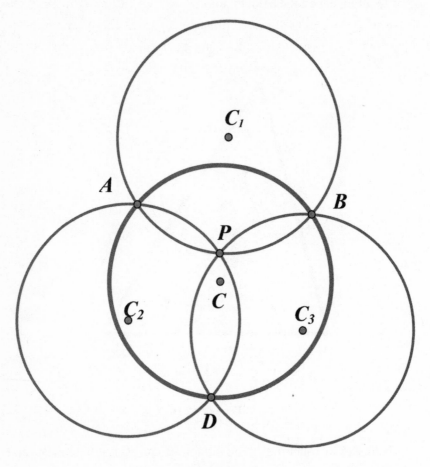

Figure 3.66

The justification for circle *C* to be the same size as the other three circles is simple to show. We begin by drawing all the radii, as shown in figure 3.67. We can easily show that three rhombuses are formed, namely, PC_2DC_3, PC_3BC_1, and PC_1AC_2. We then construct another rhombus CBC_1A.

From the various rhombuses, we can see that C_2D is equal and parallel to PC_3 and BC_1. However, CA is equal and parallel to BC_1. Therefore, CA is equal and parallel to C_2D. This then establishes that quadrilateral $ACDC_2$ is also a rhombus. That will allow us to establish that $CD = r$. We have, therefore, established that the three segments from point C to points A, B, and D are equal, which indicates that C is the center of a circle containing the points A, B, and D.

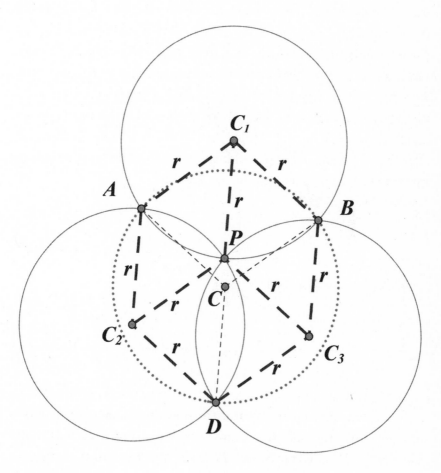

Figure 3.67

We have merely scratched the surface of the topic of concurrency, but we hope that we have motivated you to seek out other forms of concurrency involving both linear figures and circles.

SIMILARITY AND THE GOLDEN RATIO

A mathematical concept that you certainly came across in school is that of similarity, which we have discussed previously in this chapter (related to the Pythagorean Theorem). But let's review: figures are considered similar if their lengths of corresponding segments are all in the same ratio. An example of a pair of similar rectangles is shown in figure 3.68. Since 4:2 is equal to 6:3, or $\frac{4}{2} = \frac{6}{3}$, the lengths of the sides are in the same ratio and the two rectangles are considered similar. All segments in the right-hand rectangle are 1.5 times as long as the corresponding segments in the one on the left. This relationship also holds true between the diagonals of the rectangles.

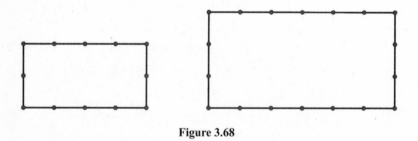

Figure 3.68

It is quite likely that you learned about similar triangles and maybe even some other similar figures. There is, however, something just a tiny step away that you may not have heard about in school but that has nevertheless fascinated people for millennia.

Consider the rectangles in figure 3.69.

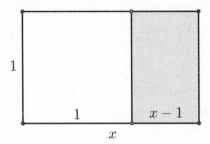

Figure 3.69

The large rectangle has sides of length 1 and x. A vertical line has been added that divides this rectangle into a square (with sides of length 1, often called a unit square) and a rectangle with sides of length 1 and $x - 1$, which is the shaded rectangle in figure 3.69. The question that has interested people since antiquity is the following: Is it possible for the small (shaded) rectangle to be similar to the original one? The answer is yes; but, as it turns out, this is only possible for one particular value of x. This value is commonly referred to as the *golden ratio* or the *golden section*, and it is usually signified by the Greek letter ϕ (phi). The numerical value of phi is approximately $\phi \approx 1.618$.

Since the ratio of the sides in the smaller (but similar) rectangle in figure 3.69 is ϕ, we can also cut off a square from this rectangle, again resulting in an even smaller similar rectangle. This can be done as often as we like, and all resulting rectangles will be similar to each other. This is shown in figure 3.70 for the first four steps.

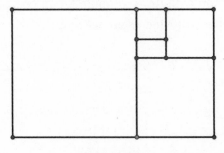

Figure 3.70

In each step, a square is cut off from the remaining rectangle, and the result is similar to all other rectangles in the figure (other than the squares, of course).

Something quite unexpected happens when quarter-circle arcs are inscribed in these squares as shown in figure 3.71. (We only show the first six of these infinitely many arcs.)

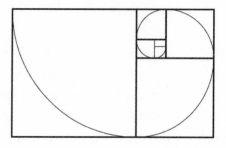

Figure 3.71

The figure composed of these quarter circles is an extraordinarily close approximation to a special logarithmic spiral known as the *golden spiral*. (If you are familiar with polar coordinates, recall that a logarithmic spiral is a curve that can be represented by the polar equation $r = ae^{b\theta}$.)

An incredible property of the golden ratio is the fact that it occurs naturally in the regular pentagon. This fact was known in antiquity, and historically this was certainly one of the reasons for humanity's longtime fascination with this ratio number. In order to see this, let us consider the regular pentagon shown in figure 3.72.

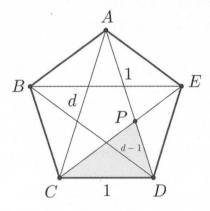

Figure 3.72

If *ABCDE* is a regular pentagon, all of its sides are sides of the same length, and we shall make that length 1 unit. Also, all diagonals of the pentagon are the same length, and we call this length *d*. Now, we can take a closer look at some of the angles in this figure. With the two diagonals *AC* and *AD*, the pentagon can be sliced in to three triangles (say *ABC*, *ACD*, and *ADE*), the sum of the five interior angles of the pentagon is equal to the sum of all the interior angles of these three triangles, or $3 \cdot 180° = 540°$. Each of these interior angles is, therefore, equal to $\frac{1}{5} \cdot 540° = 108°$. Since triangle *ABC* is isosceles ($AB = BC$), we have $\angle BCA = \frac{1}{2}(180° - \angle ABC) = 36°$, and, therefore, $\angle ACD = 108° - 36° = 72°$. We see that the triangle *ACD*, which is also isosceles ($AC = AD$), has a vertex angle of measure 36° at *A* and base angles of measure 72° at *C* and *D*, respectively.

Now, we let *P* denote the intersection point of diagonals *AD* and *CE* and consider the triangle *CDP*. $\angle PDC = 72°$, as we already know. Furthermore, $\angle PCD = 36°$, as triangles *ABC* and *CDE* are congruent. We can see that triangles *ACD* and *CDP* have two angles in common, therefore, they are similar, with $\angle DCP = 36°$. Triangle *CDP* is therefore isosceles, and we have $CP = CD = 1$. Furthermore, since $\angle ACP = 72° - 36° = 36°$, and we already know that $\angle CDP = 36°$, the triangle *ACP* is also isosceles, and we therefore also have $PA = CP = 1$.

Now let us consider the lengths of corresponding sides in triangles ACD and CDP. Due to the similarity of the triangles, we have the following: $\dfrac{AC}{CD} = \dfrac{CD}{DP}$, or $\dfrac{d}{1} = \dfrac{1}{d-1}$.

Looking back to figure 3.69, we recall that this was precisely the definition of the golden ratio, as the similar rectangles there had the proportion $\dfrac{x}{1} = \dfrac{1}{x-1}$. The lengths of the diagonals and the sides in a regular pentagon, therefore, indeed yield the golden ratio. We note that this relationship makes it quite straightforward to calculate the value of ϕ. Since $\dfrac{\phi}{1} = \dfrac{1}{\phi-1}$ is equivalent to $\phi\,(\phi - 1) = 1$, or $\phi^2 - \phi - 1 = 0$, we only need to solve this quadratic equation, and the positive solution of this is equal to

$$\frac{1}{2} + \sqrt{\frac{1}{4} + 1} \approx 1.618.$$

A RELATION BETWEEN POINTS AND CIRCLES

In plane geometry, points are the most basic objects we can think of. In some sense, they are also the most boring ones. Other geometrical objects such as straight lines, triangles, circles, and so on have certain properties that we can ascribe to them (or by which we can define them). A straight line is "straight," a triangle has three angles, a circle is "round," and so on, but a point is just a point—at least, that is, if we don't use a coordinate system. By itself, an isolated point has no properties at all, except that of being there or not. Points in geometry always need other points to acquire meaning. Remember that a straight line, for instance, is a collection of infinitely many points (and the same is true for a triangle or a circle). Only in relation to other points, or collections of points, can we talk about properties of a point, for example the distance of one point from another point. A property not so well known is the "power of a point" with respect to a circle, the latter functioning as the "other points." It is basically a real number that can be assigned to any point in the plane with respect to a given circle.

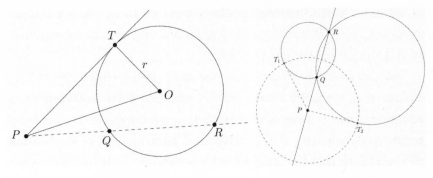

Figure 3.73   Figure 3.74

The so-called power of a point is a concept that was first defined by the Swiss mathematician Jacob Steiner (1796–1863). He wrote his publications in German and called it "*Potenz des Puncts*," which is then translated to "power of a point" (with respect to a circle). As shown in figure 3.73, when given a circle of radius r and center O, the power of a point P (p) with respect to this circle is defined as the square of the distance between the points P and O, minus the square of the radius of the circle, that is, $p = PO^2 - r^2$. As this notion involves both the circle and the point, it is, therefore, sometimes also called the "circle-power" of the point P.

Looking at figure 3.73, we can immediately see that the power of P will be positive whenever P lies outside the circle, and it will be negative whenever P lies inside the circle. The point P lies exactly on the circle if and only if $p = 0$. So, the sign of p tells us something about the relative positions of the point and the circle, whereas the absolute value of p is a measure for the distance between them. There is also a direct geometrical interpretation of p, that is, if the point P lies outside the circle, as we see in figure 3.73, then we can use the Pythagorean theorem to obtain $p = PO^2 - r^2 = PT^2$, where T is the point of tangency from P to the circle. Remarkably, for any line through P intersecting the circle in points Q and R (see figure 3.74), the product $PQ \cdot PR$ will be the same and will be equal to p. A proof of this statement can be found in the unit "When Intersecting Lines Meet a Circle." So we actually have $p = PO^2 - r^2 = PT^2 = PQ \cdot PR$ for any line through P and intersecting the circle at points Q and R.

We can then ask, What use can the power of a point be for us? Let's explore this. Consider two circles of arbitrary diameters intersecting at points Q and R. Now draw a straight line through these intersection points, as shown in figure 3.74. Then each point on line QR has the same power with respect to both circles. How can we establish this relationship? Take an arbitrary point P on this line. Then the product $PQ \cdot PR$ will be its power with respect to each of the circles. Since $PQ \cdot PR = PT^2$, this also means that all tangents drawn from P to any of the two circles have equal length. The line QR is called the *radical axis* (or "power line") of the two circles. It is the locus of points at which tangents drawn to both circles have the same length (or the locus of points having the same power with respect to both circles). The radical axis exists for any two circles. If the circles intersect, then it is the line passing through their intersection points. If the circles are tangent, it is the common tangent of the two circles. For each point on the radical axis, there is a unique circle centered on that point and intersecting the two given circles at right angles (see figure 3.74). The intersection points are the points of tangency T_1 and T_2. The converse is also true; that is, the center of a circle intersecting both given circles at right angles must lie on the radical axis. An application of this relationship can be found in the Apollonian circles, which were discovered by the Greek geometer Apollonius of Perga (ca. 262 BCE–ca. 190 BCE).

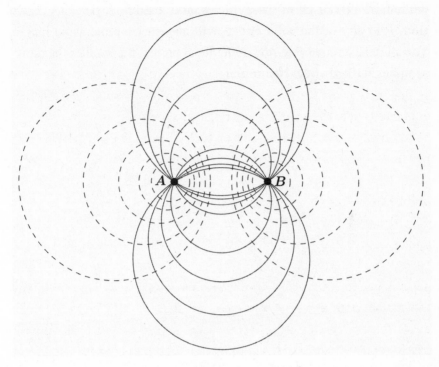

Figure 3.75

The Apollonian circles, depicted in figure 3.75, are two families of circles, for which every circle in the first family intersects every circle in the second family at right angles, and vice versa. Each pair of circles in the first family intersects in the so-called focal points A and B. Thus, as shown in figure 3.76, they all have the same radical axis AB and their centers lie on the perpendicular bisector of the line segment AB. (The centers of the four upper circles are indicated by dots.) The circles in the second family (shown dashed in figure 3.76) have their centers on the line AB. To construct such a circle, take any point P on the line through A and B, but outside the segment AB. From this point P, draw the tangent to one of the solid circles and denote the point of tangency as T. Finally, draw a circle with center P and radius PT. This circle will intersect all solid circles at right angles. This is because P lies on the radical axis of all of those circles, and, therefore, tangents drawn from P to any of the solid circles will always have the same length. The closer we move P to one of the focal points, the smaller the corresponding dashed circle becomes.

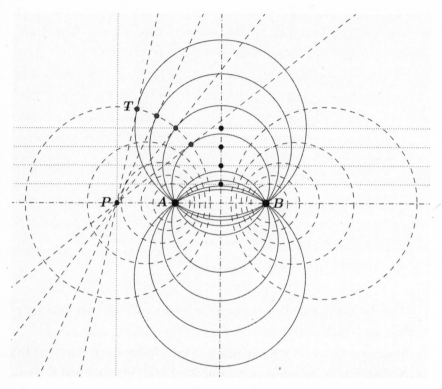

Figure 3.76

Apollonian circles can be used as an alternative way to locate a point in the plane. Usually, a point in the plane is specified by its x-coordinate and its y-coordinate, that is, its position with respect to two perpendicular coordinate axes. However, instead of using two perpendicular axes to define coordinates in the plane, one can also use other coordinate systems. For instance, a point on the surface of the earth is specified by its longitude and latitude. These are coordinates defined by lines forming circles around the earth. For each point on the surface of the earth, there is a unique circle of longitude and a unique circle of latitude running through this point. Similarly, Apollonian circles can be used to assign coordinates to points in the plane. Given the two focal points of an Apollonian system of circles, then for each point in the plane there are two unique Apollonian circles running through this point (one from each family). By specifying these two circles, for example, by means of their radii, the position of the point is uniquely determined. How might these be used outside of the classroom? Coordinates defined by Apollonian circles are very useful in describing the electric and magnetic field surrounding two parallel conductors. If A and B represent the two conductors (current flowing perpendicular to this sheet of paper and in opposite directions), then the electric field lines basically look like the family of solid circles shown in figure 3.76, whereas the magnetic field lines resemble the family of dashed circles. Unfortunately, this mathematical gem with real-world implications is often beyond the scope of the average high school mathematics class.

CONSTRUCTIONS WITH COMPASSES ALONE

A common topic in the high school geometry course is to do constructions of geometric figures using an unmarked straight edge and a pair of compasses. This allows us to draw circles and straight lines, which then enables us to do five principal constructions on which in combination all others are based. These are:

- to construct a straight line
- to construct a circle
- to construct the intersection of two lines
- to construct the intersection of two circles
- to construct the intersection of a line and a circle

In 1797 the Italian mathematician Lorenzo Mascheroni (1750–1800), a professor of mathematics at the University of Pavia, Italy, published a book called *Geometria del Compasso*. In this book, surprisingly, Mascheroni proved that all constructions, which formerly required both an unmarked straightedge and a pair of compasses, can actually be done by using compasses alone. These types of constructions are today referred to as *Mascheroni constructions*.

Curiously, in 1928, mathematicians felt a bit awkward referring to these constructions as Mascheroni constructions, since in that year a Danish mathematician, Johannes Hjelmslev, discovered a book written in 1672 by a countryman of his, George Mohr (1640–1697), who was a rather obscure mathematician, which contained arguments similar to Mascheroni's. Yet, since it is felt that Mascheroni did arrive at his conclusions independently, his name continues to be used today to identify these compasses-alone constructions.

You may be wondering how you can draw a straight line with merely the assistance of a pair of compasses. Since we know that a line consists of many points, we can show how, by using compasses alone, you can determine as many points on a given line as you need. In other words, although you would not see a continuous straight line, you have a set of points, all of which are collinear and are in a relation to one another as you would determine them to be. If you are interested in pursuing this topic further, please see *The Circle: A Mathematical Exploration Beyond the Line*, by A. S. Posamentier and R. Geretschläger (Amherst, NY: Prometheus Books, 2016).

THE SPHERE AND THE CYLINDER

Although the volume and the surface area of a sphere are usually presented as part of our geometric training in high school, the relationship of the sphere's area and volume to those of a cylinder are typically omitted. We attribute to the famous Greek mathematician Archimedes two relationships that follow from having a sphere completely contained within a cylinder so that it is tangent to the sides and top and bottom of the cylinder, as shown in figure 3.77.

Figure 3.77

We begin by comparing the surface area of the sphere to the lateral surface area of the cylinder (that is, not counting the bases of the cylinder). We find the surface area of the sphere with a radius of length r is $4\pi r^2$. Now we must determine the lateral area of the cylinder to which the sphere is tangent. The lateral area of the cylinder is a product of circumference of the base, $2\pi r$, and its height, $2r$, which is curiously also equal to $4\pi r^2$.

Let us now make an analogous comparison of the volume of the sphere to that of the cylinder around it. We know that the volume of the sphere with radius r is $\frac{4}{3}\pi r^3$. By looking at figure 3.77, we can clearly see that the volume of the sphere must be less than volume of the cylinder. We will show that the volume of the sphere is actually two-thirds that of the cylinder. The volume of the cylinder is obtained by taking the product of the area of its base and its height, which in this case is $(\pi r^2)(2r) = 2\pi r^3$. By taking $\frac{2}{3}(2\pi r^3) = \frac{4}{3}\pi r^3$, we can conclude that the volume of the sphere, $\frac{4}{3}\pi r^3$, is two-thirds of $2\pi r^3$.

There is also a rather fun comparison that can be made between the lateral area of a cylinder, whose height is half the length of the radius of the base $\left(h = \frac{1}{2}r\right)$, with the area of the base (πr^2). They are, in fact, equal. This is rather simple to show, as the area of the circular base with radius r is πr^2, while the lateral area of the cylinder, which is equal to the product of its height, $\dfrac{r}{2}$, and the circumference of its base, $2\pi r$, is also πr^2 (that is, $\dfrac{r}{2}\left(2\pi r\right)=\pi r^2$). Although this cannot be considered a proper construction in Euclidean terms, if we cut the cylinder's lateral surface along its height and spread it out to get a rectangle, we would find that the rectangle's area is equal to the circular area of the base of the original cylinder. Once again, we have a rectilinear figure equal in area to a circle. These are curious facts that often are missed during the initial academic introduction of these three-dimensional figures.

REGULAR POLYGONS AND STARS

A very basic concept you certainly came across in your school days is that of the *regular polygon*. This is, of course, a very specific type of classical figure, a closed shape composed of line segments of equal length. In order to qualify as "regular" in the usual sense, along with some other restrictions, any pair of adjacent segments must contain angles of equal size.

Typical examples of regular polygons are the equilateral triangle, the square, and the regular hexagon, as shown in figure 3.78.

Figure 3.78

Another shape that almost seems to fit the bill is one that can be found on many flags and elsewhere: the regular five-pointed star, or regular *pentagram*, which is shown in figure. 3.79.

Figure 3.79

Taking a look at this shape, we see that it fulfills all the requirements of a regular polygon stated so far. It is made up of five line segments of equal length, and any pair of these meeting in one of the points of the star contain angles of equal size (36°).

Usually, a figure like the regular pentagram is not counted among the regular polygons, though. Unlike the sides of a square or a regular hexagon, the sides of a pentagram intersect in points other than their endpoints. In order for a polygon with equal sides and equal angles to qualify as *regular* in the usual sense, we also require that it be *convex*.

It is not too easy to give a short definition of the general meaning of convex, but in the context of polygons it is perhaps easiest to think of it in the following way. Consider a line on which a side of a given polygon lies. This line divides the (infinite) plane into two halves. If the polygon lies wholly in one of these half-planes formed by the line placed on each side of the polygon, we call the polygon *convex*. An example of such a polygon is shown in figure 3.80a.

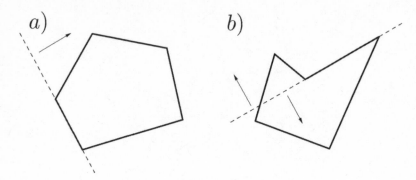

Figure 3.80

On the other hand, if parts of the polygon lie on either side of the line, or, in other words, if parts lie in both of the half-planes (as illustrated in figure 3.80b), the polygon is not convex. The latter is clearly the case for the regular pentagram, as we can see in figure 3.81.

Figure 3.81

Taking a closer look at the regular pentagram, we see that its vertices are the vertices of a regular pentagon (figure 3.82). This observation allows us to find other star polygons with similar properties.

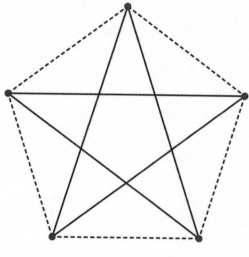

Figure 3.82

Starting with any regular polygon (as we do with a regular heptagon in figure 3.83), we can create a regular star polygon by joining nonadjacent vertices by line segments in a regular manner.

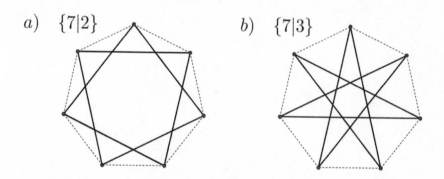

a) $\{7|2\}$ *b)* $\{7|3\}$

Figure 3.83

In figure 3.83a, we join each vertex with the next one, but skip one each time. In figure 3.83b, we join each vertex with the next one, but skip two each time. The notation generally used for such stars derived from regular *n*-gons by joining the *k*-th vertices is {*n*|*k*}. The star in figure 3.83a is {7|2}, since we have joined every other (that is, every second) vertex of a regular 7-gon. Similarly, the star in figure 3.83b is {7|3}, since we have joined every third vertex of a regular 7-gon. In this notation, the pentagram in figure 3.82 is, of course, {5|2}, and any regular *n*-gon is {*n*|1}.

We can now use this notation as an easy way to write ideas concerning what stars of this type exist. On the one hand, we have just noted that {*n*|1} denotes a regular *n*-gon. On the other hand, {*n*|*n* − 1} denotes a regular figure in which we join any vertex of a regular *n*-gon to the one *n* − 1 vertices on. Going *n* − 1 vertices to the left in an *n*-gon, say, is just the same as going 1 vertex to the right, however. In fact, we see that {*n*|1} and {*n*|*n* − 1} refer to the same object. By the same reasoning, the expressions {*n*|*k*} and {*n*|*n* − *k*} refer to the same object for all values of *k* with $1 \le k \le n - 1$. In general, we can assume $1 < k \le \frac{n}{2}$ anytime we use this notation, since this will certainly let us refer to any regular star.

We now note that there are values of *k* we can choose that will not yield stars of a closed type, as we had with the pentagram. For instance {6|3}, shown in figure 3.84a, gives us an asterisk-like object; a "star" that has split into three line segments. On the other hand, {6|2}, shown in figure 3.84b, gives us a star composed of two equilateral triangles.

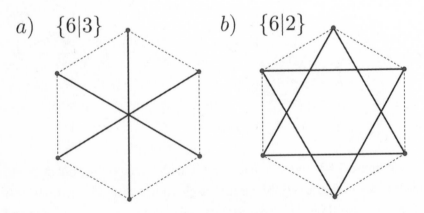

Figure 3.84

Under some circumstances, this concept of a "star" might be just what we want, but often we prefer to think of a star shape as being a single closed polyline, that is, a figure made up of line segments that can be drawn without lifting the pencil from the paper and ending at the same point from which we started. This is only true for $\{n|k\}$ if n and k have no common divisor greater than 1, that is to say, if they are relatively prime.

Using this restriction, we see that the question of finding out how many such n-pointed stars exist for a given n is equivalent to the question of how many integers k exist, not greater than $\frac{n}{2}$ and relatively prime with n. The combinatorial question of "how many" in the geometric context of "stars" remarkably turns out to be solvable by a number-theoretical approach.

Some more examples of such stars are shown in figure 3.85.

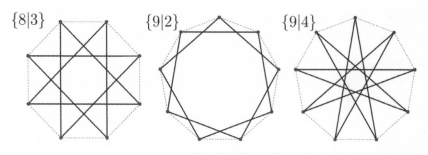

Figure 3.85

In fact, there exists only this one closed (polyline) star $\{8|3\}$ for $n = 8$, and only these two $\{9|2\}$ and $\{9|4\}$ for $n = 9$. You might like to try your hand at drawing the one possible star for $n = 10$ and the four stars that exist for $n = 11$ next.

PLATONIC SOLIDS AND STAR POLYHEDRA

In the last section, we saw how a slight loosening of the strict definition of a regular polygon could lead us to the closely related star polygons. Now, we can take a step up by one dimension, and consider what
happens when we expand this idea into the three-dimensional world of
solid geometry.

Just as the stars in the plane result by joining nonconsecutive vertices of regular polygons in a systematic way (recall how we could
create the sides of the regular pentagram by joining every other vertex
of a regular pentagon), something similarly wonderful results when we
join the vertices of the Platonic solids in such a way.

Perhaps you will recall from your school days that the Platonic
solids are the most regular of all the polyhedra. You will certainly recognize them in figure 3.86; in order from left to right, the tetrahedron,
the hexahedron (or cube), the octahedron, the dodecahedron, and the
icosahedron. Each is, of course, named for the Greek word for the
number of its faces: four, six, eight, twelve, and twenty, respectively.

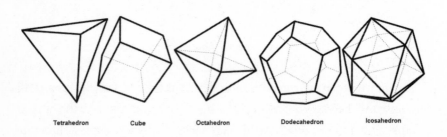

Tetrahedron   Cube   Octahedron   Dodecahedron   Icosahedron

Figure 3.86

Each of these solids has faces that are all congruent regular poly-gons. Similar to the defining properties of the regular polygons, each pair of faces contains an angle of equal size. And, as was the case for regular polygons, we also require that the solids be convex. (In three-dimensional space, this can be seen as meaning that the entire solid lies on one side of any of the planes containing a face, as suggested in figure 3.87.)

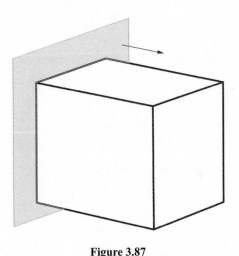

Figure 3.87

Now let's consider what will happen if we try to create a new solid by starting with the vertices of a Platonic solid and then join nonadja-cent pairs of these vertices to create the edges of a new solid. Nothing of particular interest happens if we try this with a tetrahedron or an octahedron. This procedure will not produce a new solid for these two polyhedra. However, joining the vertices of a hexahedron (the six-sided Platonic solid you may be more accustomed to thinking of as a cube) yields the famous *stella octangula*. This is Latin for "eight-pointed star," and it is depicted in figure 3.88.

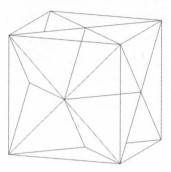

Figure 3.88

Each diagonal on the side (or face) of the cube is an edge of the stella octangula, but there are also some additional edges. The reason for this is as follows. This solid—sometimes attributed to the German astronomer and mathematician Johannes Kepler (1571–1630)—can be thought of as the union of two regular tetrahedra, as we see in figure 3.89. In this sense, the stella octangula is not really a "new" solid.

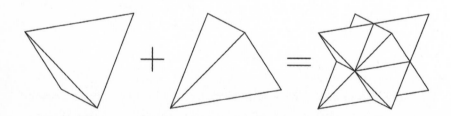

Figure 3.89
The two tetrahedra intersect in additional lines that are then also edges of the stella octangula. These do not join vertices of the original cube, but rather join midpoints of its faces. For this reason, the stella octangula is not usually considered one of the "regular" star polyhedra.

Joining the vertices of the more complex dodecahedron or icosahedron in similar ways does, however, yield completely new star-shaped polyhedra, all of whose edges connect vertices of the Platonic solid. These are commonly referred to as the *Kepler-Poinsot solids*. These were named after Kepler and the French mathematician and physicist Louis Poinsot (1777–1859), who first wrote extensively about them in the mathematical literature.

Kepler wrote about the small stellated dodecahedron (figure 3.90)[8] and the great stellated dodecahedron (figure 3.91),[9] and Poinsot completed the set two centuries later with the great dodecahedron (figure 3.92)[10] and the great icosahedron (figure 3.93).[11] The vertices of the great stellated dodecahedron are also the vertices of a regular dodecahedron, while those of each of the other three are also the vertices of a regular icosahedron.

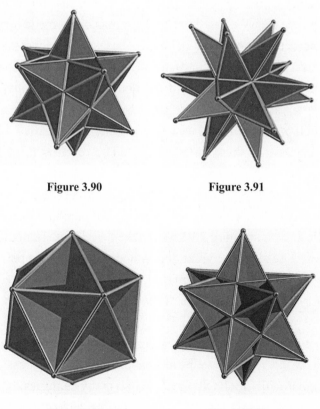

Figure 3.90   Figure 3.91

Figure 3.92   Figure 3.93

Figure 3.94

In figure 3.94, we see three different ways in which we can think of creating the small stellated dodecahedron. In figure 3.94a, we see the edges of the icosahedron that shares vertices with the star. In figure 3.94b, we see a small pyramid (on the left side) that forms one of the "points" of the star. We can think of the star as being made up of twelve such five-sided pyramids that have their bases on the faces of a regular (pentagonal) dodecahedron in such a way that the edges of the pyramids are continuations of the edges of the dodecahedron. Finally, in figure 3.94c, we see a pentagram. We can also think of the star as having twelve such pentagrams as faces. (In fact, the star is called a "dodecahedron" because of these twelve faces.) The result is not convex, of course, as these faces intersect one another.

Similar to what we see in figure 3.94c, the great stellated dodecahedron also has twelve pentagrams as faces, but combined in a different way. A close look at figure 3.91 will make these pentagrams visible for the interested reader. The great dodecahedron has twelve pentagons as (intersecting) faces, and the great icosahedron has twenty equilateral triangles as (intersecting) faces. These, too, can be found by taking a closer look at figures 3.94 and 3.93. There is certainly a lot of fascinating symmetry to be found in the interior of any of these star polyhedra. No wonder they have been a favorite topic of discussion for so many geometrically inclined people for so many centuries.

The field of geometry offer so much more than a school curriculum can accommodate. Many of these outliers are extremely interesting yet are often based on a reasonably good foundation of knowledge composed by the basics taught within the school curriculum. There are many sources where such topics can be found, and we hope that you will seek them out (we refer you especially to other works by these authors, some of which are listed at the front of this book). We now move to another field of mathematics that only recently has been given a foothold in the curriculum, but not to the extent that some of the gems we will present have yet found a place.

CHAPTER 4

PROBABILITY APPLIED TO EVERYDAY EXPERIENCES

A mathematical topic that is increasingly working its way into the curriculum is that of probability. Looking back fifty or more years, this topic was, at best, relegated to the senior year in high school, and even there it was hardly visible. Many speculate why that might have been the case. Perhaps because this topic requires more mature, sophisticated thinking? In any case, there is quite a bit of material that could have been presented and that would have enhanced the mathematical experience of the students at the time. We will begin by mentioning a brief rendering of the initiation of this important field of mathematics, and then we will delve into some common applications, a number of which will be quite surprising—yet true!

HOW THE THEORY OF PROBABILITY BEGAN

As one would expect, it was gambling that sparked the beginning of probability theory. In the seventeenth century, two famous French mathematicians Blaise Pascal (1623–1662) and Pierre de Fermat (1601–1665) were tangling with a problem involving the flipping of a coin. The game that they were considering was to be played as follows: if the coin comes up heads, then player A gets a point; if the coin comes up tails, then player B gets a point. For this game, the first player to reach ten points is the winner. In other words, if player A gets ten points before player B, then player A is the winner. The problem that they grappled with was, if this game gets interrupted before a winner is determined, how should the money being wagered be split among the two participants?

Clearly, the player who has more points should get more money. However, in what ratio should the money be distributed among the two players? Would splitting the money according to the ratio of points earned be fair? Consider what would happen if the score was $1 - 0$. That would indicate that one player would get all the money, and the other player would get nothing, even though the difference between the scores is minimal. This does not seem fair.

Rather than to look at scores that were achieved at the point of the game's disruption, one might look at the number of points each one still needed to get to reach the required winning score of 10. In other words, the money might be divided in the ratio of their probabilities of winning, were the game to be completed after the pause.

Suppose the game were interrupted at the point where player A had 7 points and player B had 9 points. There would be at most three more flips of the coin necessary to determine the winner. These flips could be any of the following: HHH, HHT, HTT, THH, TTH, TTT, HTH, THT.

In order for player A to win, he would have to flip HHH, that is, only one of the eight possibilities, or a probability of $\frac{1}{8}$, while player B could win with any of the remaining possibilities, or a probability of $\frac{7}{8}$. Therefore, Pascal and Fermat concluded that the money at that point in the disrupted game should be split up in the ratio of 1:7.

This is one of the first problems that generated the field of probability. It gives us some understanding of the kind of thinking that goes into solving problems involving probabilistic thinking.

BENFORD'S LAW

When you were studying elementary probability in school, you were almost certainly confronted with the concept of a *discrete uniform distribution*. This term is used to describe an experiment or observation with a (finite) number of possible outcomes, each of which is equally likely to happen. For a slightly more mathematical way to put it, we could also say that the probability of each of the possible outcomes is equal.

Distributions we come across in most situations tend not to be uniform, even when we expect them to be. In fact, this is true for real-life data as well as many purely theoretical collections of numbers. The work of the American physicist Frank Albert Benford Jr. (1883–1948) explained this surprising fact as what is known as *Benford's law*. Before we get into this, however, let us recall what we know about distributions that are uniform.

A very typical example of this is flipping a coin. The likelihood of heads or tails coming up is 50/50, or in the language of mathematics, we can say the probability of either heads or tails is equal to $\frac{1}{2}$. Other examples are rolling a die, in which case the probability of any of the numbers from 1 through 6 coming up is equal to $\frac{1}{6}$; or spinning a roulette wheel, for which the probability of the little ball landing on any one of the 37 numbers from 0 through 36 is equal to $\frac{1}{37}$ (or $\frac{1}{38}$, if you are playing with a wheel that also includes double zero).

It is not surprising that the archetypical examples of this concept come from the world of gambling. After all, we speak of "games of chance" when all outcomes should be equally likely, with no skill or manipulation involved, at least ideally. All of modern probability theory, with its myriad applications in the worlds of finance, medicine, politics, and others, can be historically traced back to such ideas.

Interestingly, it is not always obvious to decide whether the distribution resulting from a specific situation is actually completely uniform. Here is where we can encounter a fascinating fact that you, in all likelihood, were not confronted with in school.

Let us consider the two-digit numbers (10 through 99), and imagine choosing one of them at random. What is the probability that some specific digit, say, 4, would be the first digit of this number? We can decide this by simple counting. There are ten numbers in the group with the first digit 1 (referring to the numbers 10 through 19, of course), 10 with the first digit 2 (20 through 29), and so on. This is also the case for 4 as a lead digit, with the ten numbers 40 through 49 having this property. This means that the probability of the randomly chosen digit starting with 4 is equal to $\frac{10}{90}$, or $\frac{1}{9}$. In this way, we see that the probability of the first digit of a randomly chosen number from this group is equal to $\frac{1}{9}$ for each of the digits 1 through 9.

Next, let us consider the same thing if we choose a number at random from the set of all three-digit numbers (100 through 999). Once again, the probability for each of the possible lead digits 1 through 9 is equal to $\frac{1}{9}$, as there are 100 three-digit numbers among the 900 that exist altogether with that specific digit in the first spot (for instance, the 100 numbers from 400 through 499 all have the lead digit 4). A similar argument holds for the set of all four-digit numbers, or for the set of all five-digit numbers, and so on.

Since this is obviously also true for the set of all one-digit numbers (each number is its own lead digit in this case), we see that the probability of choosing a number with any specific lead digit from 1 through 9 out of the set of all numbers (positive integers, of course) with at most a specific number of digits (for instance, if we take the set of all numbers with at most four digits, they can have 1, 2, 3, or 4 digits), is always $\frac{1}{9}$. This is, therefore, a discrete uniform distribution (that is, a symmetric probability distribution).

An easy (but *incorrect*) jump to make is now to say that the probability of a specific lead digit is always evenly distributed and, therefore, at least approximately equal to $\frac{1}{9}$ in any sufficiently large set of numbers. While a strong argument can be made for the validity of this if we are choosing from the set of all positive integers, it is incorrect if we are choosing numbers from some limited set, even if the set is very large. An easy example of a set in which this is incorrect would be the set of page numbers of a book. Let us assume that we have a book with 200 pages numbered consecutively from 1 to 200. It is immediately obvious that more than half of the page numbers of this book start with the digit 1, and so the probability of the lead digit being 1 is greater than $\frac{1}{2}$, and certainly much larger than $\frac{1}{9}$. In fact, in this case, 111 page numbers start with the number 1; 12 pages start with the number 2; and 11 start with each of the other digits (excluding 0, of course). In this case, the probability distribution is anything but uniform.

We might assume that the probabilities may still be quite close to equal if the sets we are choosing from are sufficiently large, but this is where Benford's law comes in. If the sets of numbers are very large and are derived from some concrete context (say phone bills, birth rates, physical constants, heights of mountains or similar things, and so on), the probabilities tend to give an advantage to the smaller digits. The probability of 1 as a leading digit tends to be about 30 percent, then the percentage falls as the digits grow, with the probability of 9 as a leading digit tending to be only about 5 percent. In fact, the probabilities come close to the values obtained from the formula $\log_{10}\left(1+\frac{1}{d}\right)$ for each of the digits $d \in \{1, 2, 3, 4, 5, 6, 7, 8, 9\}$. Interestingly, the law also holds for many sets of theoretically produced numbers, such as powers of 2, the Fibonacci numbers, or factorials.

A complete argument for this phenomenon is not easy to provide, but the basic idea has a lot to do with the example of the page numbers mentioned above. When we count, the successive numbers resulting tend to use all the digits equally often, but the lead digits are an exception. Of course, zero can never be the lead digit, so there you immediately have a difference compared to the other digits. Furthermore, when we count, the lead digits only change when there is a carryover from the next-to-last digit (as in 27, 28, 29, 30, 31, 32, . . .), and any time such a carryover creates a new digit, this will be 1 for quite a while. For instance, after counting the first 999 numbers and reaching 997, 998, 999, 1000, . . . the next 1,000 numbers all start with the digit 1.

Benford's law turns out to be quite useful in deciding how "real" a large set of numbers, which appears in some context, is. This is particularly useful to differentiate computer generated random number lists from lists of things like telephone numbers, Social Security numbers, or other account numbers. A simple count of the digits will show whether they are evenly distributed or whether they appear in accordance with the expectations of Benford's law. Using this law, strong arguments have, for instance, been brought forward to give evidence of things like election fraud (a commonly cited example is the Iranian election of 2009) or falsified economic data (an example often noted in this connection being the data given by the Greek government in order to join the Eurozone). Of course, such arguments are not conclusive, but they can certainly be used effectively to harden suspicions of falsified data.

THE BIRTHDAY PHENOMENON

Fortunately, the topic of probability is becoming ever more popular in the school curriculum today. Most of the results are intuitively logical (e.g., the probability of flipping a coin and getting a head is one half; the probability of throwing a die to get a 2 is one-sixth; and so on). We also know that the probability of winning a lottery is rather small, but yet there are some results in the field of probability that are truly counterintuitive. Here we will present a situation that could be described as one of the most surprising results in mathematics. It is one of the best ways to convince the uninitiated of the "power" of probability. We hope with this example not to upset your sense of intuition.

Let us suppose you are in a room with about thirty-five people. What do you think the chances are (or the probability is) that two of these people have the same birth date (month and day only). Intuitively, you usually begin to think about the likelihood of 2 people having the same date out of a selection of 365 days (assuming no leap year). Perhaps 2 out of 365? That would be a probability of $\frac{2}{365} = .005479 \approx \frac{1}{2}\%$. A minuscule chance.

Let's consider the "randomly" selected group of the first thirty-five presidents of the United States. You may be astonished that there are two with the same birth date: the eleventh president, James K. Polk (November 2, 1795), and the twenty-ninth president, Warren G. Harding (November 2, 1865).

You may be surprised to learn that for a group of thirty-five, the probability that two members will have the same birth date is greater than 8 out of 10, or $\frac{8}{10} = 80\%$.

If you have the opportunity, you may wish to try your own experiment by selecting ten groups of about thirty-five members to check on date matches. For groups of thirty, the probability that there will be a match is greater than 7 out of 10, or, in other words, in 7 of these 10 groups there ought to be a match of birth dates. What causes this incredible and unanticipated result? Can this be true? It seems to be counterintuitive.

To relieve you of your curiosity, we will investigate the situation mathematically. Let's consider a class of thirty-five students. What do you think is the probability that one selected student matches his own birth date? Clearly *certainty*, or 1. This can be written as $\frac{365}{365}$.

The probability that another student does *not* match the first student is $\frac{365-1}{365} = \frac{364}{365}$.

The probability that a third student does *not* match the first and second students is $\frac{365-2}{365} = \frac{363}{365}$.

The probability of all thirty-five students *not* having the same birth date is the product of these probabilities: $p = \frac{365}{365} \cdot \frac{365-1}{365} \cdot \frac{365-2}{365} \cdots \frac{365-34}{365}$.

Since the probability (q) that two students in the group *have* the same birth date and the probability (p) that two students in the group *do not have* the same birth date is a certainty, the sum of those probabilities must be 1. Thus, $p + q = 1$.

214 THE JOY OF MATHEMATICS

In this case,

$$q = 1 - \frac{365}{365} \cdot \frac{365-1}{365} \cdot \frac{365-2}{365} \cdots \cdots \frac{365-33}{365} \cdot \frac{365-34}{365} \approx .8143832388747152 .$$

In other words, the probability that there will be a birth date match in a randomly selected group of thirty-five people is somewhat greater than $\frac{8}{10}$. This is quite unexpected when you consider that there were 365 dates from which to choose. If you are feeling motivated, you may want to investigate the nature of the probability function. Here are a few values to serve as a guide:

Number of People in Group	Probability of a Birth Date Match
10	.1169481777110776
15	.2529013197636863
20	.4114383835805799
25	.5686997039694639
30	.7063162427192686
35	.8143832388747152
40	.891231809817949
45	.9409758994657749
50	.9703735795779884
55	.9862622888164461
60	.994122660865348
65	.9976831073124921
70	.9991595759651571

Notice how quickly "almost-certainty" is reached. With about 55 students in a room the chart indicates that it is almost certain (.99) that two students will have the same birth date.

Were you to do this with the death dates of the first thirty-five presidents, you would notice that two died on March 8 (Millard Fillmore in 1874 and William H. Taft in 1930) and three presidents died on July 4 (John Adams and Thomas Jefferson in 1826, and James Monroe in

1831). This latter case leads some people to argue that you could pos-sibly will you own date of death, since these three presidents died on their most revered day in American history! Above all, this astonishing demonstration should serve as an eye-opener about the inadvisability of relying too much on your intuition.

THE MONTY HALL PROBLEM

While the previous unit showed us that some probability results are quite counterintuitive, here we will show a very controversial issue in probability that also challenges our intuition. There is a rather famous problem in the field of probability that is typically not mentioned in the school curriculum, yet it has been very strongly popularized in newspa-pers, magazines and even has at least one book[1] entirely devoted to the subject. It is one of these counterintuitive examples that gives a deeper meaning to understanding the concept of probability.

This example stems from the long-running television game show *Let's Make a Deal*, which featured a rather curious problematic situ-ation. Let's look at the game show in a simplified fashion. As part of the game show, a randomly selected audience member would come on stage and be presented with three doors. She was asked to select one, with the hope of selecting the door that had a car behind it, and not one of the other two doors, each of which had a donkey behind it. Selecting the door with the car allowed the contestant to win the car. There was, however, an extra feature in this selection process. After the contes-tant made her initial selection, the host, Monty Hall, exposed one of the two donkeys, which was behind a not-selected door—leaving two doors still unopened, the door chosen by the contestant and one other door. The audience participant was asked if she wanted to stay with her original selection (not yet revealed) or switch to the other unopened door. At this point, to heighten the suspense, the rest of the audience would shout out "stay" or "switch" with seemingly equal frequency. The question is, what to do? Does it make a difference? If so, which is the better strategy to use here (i.e., which offers the greater probability of winning)? Intuitively, most would say that it doesn't make any difference, since there are two doors still unopened, one of which conceals a car and the other a donkey. Therefore, many folks would assume there is a 50/50 chance that the door the contestant initially selected is just as likely to have car behind it as the other unopened door.

Let us look at this entire situation as a step-by-step process, and then the correct response should become clear. There are *two donkeys* and *one car* behind these three doors. The contestant must try to get the car. Let's assume that she selects Door 3. Using simple probabilistic thinking, we know the probability that the car is behind Door 3 is $\frac{1}{3}$. Therefore, the probability that the car is behind either Door 1 or Door 2 is then $\frac{2}{3}$. This is important to remember as we move along.

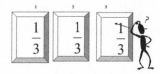

Figure 4.1

Knowing where the car is hidden, host Monty Hall then opens one of the two doors that the contestant did *not* select, and exposes a donkey. Let's say that the contestant chose Door 3 and Monty revealed Door 2 (with a donkey). Keep in mind that the probability that the car is behind one of these two remaining doors, Doors 1 and 2, is $\frac{2}{3}$.

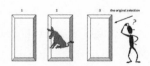

Figure 4.2

He then asks the contestant, "Do you still want your first-choice door, or do you want to switch to the other closed door?" Remember, as we said above, the combined probability of the car being behind Door 1 or Door 2 is $\frac{2}{3}$. Now with Door 2 exposed as not having the car behind it, the probability that the car is behind Door 1 is still $\frac{2}{3}$, while we recall the probability that the car is behind Door 3, the door initially selected by the contestant, is still only $\frac{1}{3}$. Therefore, the logical decision for the contestant is to switch to Door 1. (Door 1's probability for having the car is $\frac{2}{3}$; Door 3's is the lesser $\frac{1}{3}$.)

This problem has caused many an argument in academic circles, and it was also a topic of discussion in the *New York Times*, as well as other popular publications. The science writer John Tierney wrote in *The New York Times* (Sunday, July 21, 1991):

> Perhaps it was only an illusion, but for a moment here it seemed that an end might be in sight to the debate raging among mathematicians, readers of *Parade* magazine, and fans of the television game show *Let's Make a Deal*. They began arguing last September after Marilyn vos Savant (1946–) published a puzzle in *Parade*. As readers of her Ask Marilyn column are reminded each week, Ms. vos Savant is listed in the *Guinness Book of World Records* Hall of Fame for "Highest I.Q.," but that credential did not impress the public when she answered this [Monty Hall probability] question from a reader.

She gave the right answer, but still many mathematicians argued.

Although this is a very entertaining and popular problem, it is extremely important to understand the message herewith imparted; and it is one that by all means should have been a part of the school curriculum to make probability not only more understandable but also more enjoyable.

BERTRAND'S BOX

If you managed to wrap your mind around the Monte Hall paradox, you might also enjoy trying your hand at the very similar (in fact, mathematically equivalent) question of Bertrand's box. This problem, named after the French mathematician Joseph Bertrand (1822–1900), was first published in 1889 and is something that should further enhance your understanding of probability.

Imagine that there are three boxes in front of you. One contains two gold coins, one contains two silver coins, and the third contains one coin of each type. You are invited to choose one of the three boxes at random, and then to take one of the coins from the chosen box without looking at it. When you place it on the table, you find that you have chosen a gold coin. What is the probability that the other coin in the chosen box is then also gold?

It seems almost too easy to be a question at all. There are equal numbers of silver coins and gold coins in play, so the situation must be completely symmetrical, right? In other words, the probability must be 50 percent, right? Well, no, that's not correct!

If you have already given some thought to the Monte Hall paradox, you might already know the answer. You will certainly be wary of jumping to any rash conclusions. In fact, the situation is not completely symmetric, as you have the information that the first coin you selected is gold. Seen in this light, it might even be the case that the probability of the second coin in the box being gold as well could conceivably be less than 50 percent. On the other hand, there are two boxes that contain a gold coin, so you know that the chosen box is one of these. One of these boxes has a second gold coin in it, and one has a silver coin in it. So is it 50 percent after all?

As it turns out, the probability that the other coin in the box is also gold is actually $\frac{2}{3}$. There are several ways to see this.

Let us make it plausible that the probability of the second coin being gold is at least higher than the probability of it being silver. The easiest way to do this is by considering what happens if you play the game very often, let's say three million times. First of all, in each game, you choose a box. Since you are equally likely to choose any of the three boxes, you will expect to choose each box about a million times. If the chosen box was the one with two silver coins, the coin you put on the table was certainly not gold. If the box was the one with two gold coins, the coin on the table was certainly gold. Finally, if you chose the "mixed" box, the coin selected will be gold half of the time. That means that the coin on the table will be gold 1.5 million times, namely, all the million times you chose the gold-gold box and half a million times out of the those in which you chose the mixed box. Of these 1.5 million times, the other coin is gold one million times, namely, all the times you chose the gold-gold box. That is one million times out of 1.5 million, or $\frac{2}{3}$ of the time.

Still not satisfied? Here is another way to better understand what is going on.

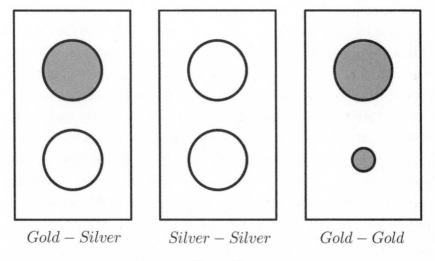

Gold – Silver   *Silver – Silver*   *Gold – Gold*

Figure 4.3

In figure 4.3, we have represented the three boxes by rectangles. The gold coins are represented by the shaded circles; and the silver coins, by white circles. Notice that the gold-gold box contains two coins of different sizes: a large gold coin and a small gold coin.

Now, let us think of the problem conditions in a slightly different way. Instead of choosing a box first and then choosing a coin at random from within that chosen box, we can just choose any of the six coins at random. After this preliminary step, we consider the probability that the chosen coin, which we know to be gold, was in the same box as another gold coin. Of course, the probability of choosing a silver coin or a gold coin is equal, if we see the setup in this way, as there are three coins of each type in play from the beginning.

We know that we have chosen a gold coin. There are three of these in play. One of these is the one in a box together with a sliver coin. One is the large gold coin in the gold-gold box, and one is the small gold coin in the gold-gold box. Only one of these three gold coins is paired with a silver coin. The other two are each paired with a gold coin (large with small and small with large), and therefore the probability of the selected gold coin being paired in a box with another gold coin is again seen to be $\frac{2}{3}$.

As a last, quite powerful method, we can use the tools at our disposal from conditional probability. If we write P(gold|gold-gold) for the probability that we select a gold coin on the table under the assumption that the coin was taken from the gold-gold box, and we write P(gold|mixed) for the probability that we select a gold coin on the table under the assumption that the coin was taken from the mixed box, and we write P(gold|silver-silver) for the probability that we select a gold coin on the table under the assumption that the coin was taken from the silver-silver box, then Bayes's rule, first presented by the English statistician Thomas Bayes (1701–1761), gives us:

$$\frac{P(\text{gold} \mid \text{gold} - \text{gold})}{P(\text{gold}|\text{gold} - \text{gold}) + P(\text{gold}|\text{mixed}) + P(\text{gold} \mid \text{silver} - \text{silver})} = \frac{1}{1 + \frac{1}{2} + 0} = \frac{2}{3}.$$

In other words, the probability that the visible gold coin was chosen from the gold-gold box, and is therefore paired with another gold coin, is once again seen to be equal to $\frac{2}{3}$. Exposure to this thinking would surely enrich your understanding of probability.

THE FALSE POSITIVE PARADOX

When you were learning about probability and statistics in school, you almost certainly had the opportunity to give a bit of thought to practical applications of the basic concepts. Of all the wonderful areas of mathematics, this is perhaps the topic we are most often confronted with in daily life. A large amount of data important to our day-to-day dealings is delivered to us in some boiled-down statistical form, whether it shows itself in the guise of pie charts in the daily paper, batting averages of the players on the local team stated in thousandths, or probabilities of something happening stated in terms of percentages (such as the likelihood of precipitation for the day).

It is perhaps not surprising that the simplification of complex data to just a few numbers would lead to some loss of information. Nevertheless, taking a closer look at some ideas we might have about such statistical information can yield some big surprises. An example of such a surprise you may not have heard about in school is the so-called *false positive paradox*. Typically, this counterintuitive consequence of the imprecise interpretation of probabilities turns up in discussions of testing for diseases, and so we will use that context here as well.

Imagine that you are going to the doctor for a general checkup, and part of this is a test for a specific disease. For the sake of this discussion, let's call it B-fever. You are told that this recently developed test is the best ever for B-fever, and it is 99 percent accurate. A week later, you get the results of your test, and you are told that you have tested positive. A typical reaction at this point (but not one supported by the numbers, as we will see), is to assume the worst. After all, the test is 99 percent accurate. So that means that there is a 99 percent chance that you have B-fever, right? Well, no. Let's take a closer look at the meaning of the numbers.

When we state that a test is 99 percent accurate, we are, in fact, not being very precise. Do we mean that 99 percent of all people with the disease will be diagnosed as having it (and therefore 1 percent as not having it even though they do, so-called *false negatives*), or do we mean that 99 percent of all healthy people will be correctly diagnosed? Or do we mean both? To simplify things a bit, we can assume here that both of these assumptions are true. (Note that for real tests of this type, these numbers will typically not be the same. The percentage of *false positives* often differs from the percentage of *false negatives*.) So we assume that 99 percent of all people taking the test get a correct result and 1 percent get an incorrect one. It seems obvious that this also means that 99 percent of all people testing positive will, in fact, be infected with B-fever.

In order to see that this is not correct, it is probably easiest to consider a hypothetical specific population and check their numbers in detail. Let us assume that we are testing 100,000 people for B-fever. We need to make some assumption on the number of people in the population that are actually infected. To this end, we assume that 0.1 percent of the total population actually has B-fever. This means that of the 100,000 people, 0.1 percent, or 100 people, have B-fever; the remaining 100,000 − 100 = 99,900 do not have B-fever. These assumptions then give us the data collected in the following table.

	Test Results Positive	Test Results Negative	**Total**
Infected with B-fever	99	1	100
Not Infected	999	98,901	99,900
Total	1,098	98,902	100,000

Of the 100 infected members of the tested population, 99 test positive and 1 tests negative, since the test is 99 percent accurate. However, of the 99,900 healthy individuals, 1 percent also tests positive. This means that 1 percent of the 99,900 people, or 999 healthy individuals, are testing positive; therefore, 99,900 − 999 = 98,901 people are testing negative. Adding these numbers, we see that 99 + 999 = 1098 people test positive. This means that, out of all individuals testing positive, only $\frac{99}{1098} \approx 9.2$ percent are actually infected with B-fever. Put another way, even if you test positive, the probability that you are not infected is still about 100 − 9.2 = 90.8 percent. This is a far cry from the 1 percent we may have naively assumed.

In fact, the less prevalent the infection is in the total population, the less likely a positive test is to imply actual infection. If only 0.01 percent of the population is infected, the percentage drops to $\frac{99}{99 + 9999} \approx 1$ percent. (These numbers result from the same idea as above, but with a tested population of one million.) We see that it is still quite unlikely that you actually have some infection, even if you test positive for it. Especially as it is a relatively rare thing. This is why more tests are needed under such circumstances. Of course, two false positives in a row become much less likely, and a second test will give us a better idea of the actual situation.

On the other hand, if you test negative, the likelihood that you are free of infection is quite high. Once again, let us consider the numbers in our test population. Of the 98,902 individuals whose tests come back negative, 98,901 are actually free of infection. The probability that someone testing negative is actually not infected is therefore equal to $\frac{98,901}{98,902} \approx 99.999$ percent. This is such a high probability that it is about as close to a sure thing as we can get by statistical methods.

The somewhat surprising behavior of these numbers is typical for conditional probabilities. Working with such values takes some getting used to, and we need to be careful not to blunder into any mental traps when dealing with such things. Here again we see that studying this surprising aspect of probability is worth teaching in high school, as it has real-world implications.

PASCAL'S TRIANGLE

Perhaps one of the most famous triangular arrangements of numbers is the Pascal triangle (named after the French mathematician Blaise Pascal, 1623–1662). Although it was likely introduced in school in connection with the binomial theorem or with probability, it has many interesting properties beyond that field, many of which could be used to enhance our appreciation of mathematics. Let's see how this triangular arrangement of numbers is constructed. Begin with a 1, then beneath it 1, 1, and then begin and end each succeeding row with a 1 and get the other numbers in the row by adding the two numbers above and to the right and left. Following this pattern so far, we would then have the following.

$$1$$
$$1\ 1$$
$$1\ 2\ 1$$
$$1\ 3\ 3\ 1$$

Continuing with this pattern the next row would be

$1 - (1 + 3) - (3 + 3) - (3 + 1) - 1$, or $1 - 4 - 6 - 4 - 1$.

A larger version of the Pascal Triangle is shown in figure 4.4.

```
                          1
                       1     1
                    1     2     1
                 1     3     3     1
              1     4     6     4     1
           1     5    10    10     5     1
        1     6    15    20    15     6     1
     1     7    21    35    35    21     7     1
  1     8    28    56    70    56    28     8     1
1     9    36    84   126   126    84    36     9     1
1   10    45   120   210   252   210   120    45    10     1
```

Figure 4.4

In probability, the Pascal triangle emerges from the following example (see figure 4.5). We will toss coins and calculate the frequency of each event, as seen in the third column.

Number of Coins	Number of Heads	Number of Arrangements
1 Coin	1 Head	1
	0 Heads	1
2 Coins	2 Heads	1
	1 Head	2
	0 Heads	1
3 Coins	3 Heads	1
	2 Heads	3
	1 Head	3
	0 Heads	1
4 Coins	4 Heads	1
	3 Heads	4
	2 Heads	6
	1 Head	4
	0 Heads	1

Figure 4.5

What makes the Pascal triangle so truly outstanding is the many fields of mathematics it touches. In particular, there are many number relationships present in the Pascal triangle. For the sheer enjoyment of it, we shall consider some here.

The sum of the numbers in the rows of the Pascal triangle are the powers of 2 (see figure 4.6).

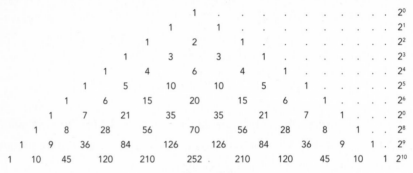

Figure 4.6

In figure 4.7, if we consider each row as a number with the members of the row their digits, such as 1, 11, 121, 1331, 14641, and so on (until we have to regroup from the sixth row on), you will find that these are the powers of 11, that is, 11^0, 11^1, 11^2, 11^3, 11^4.

					1						11^0
				1		1					11^1
			1		2		1				11^2
		1		3		3		1			11^3
	1		4		6		4		1		11^4
1		5		10		10		5		1	11^5
1	6		15		20		15		6	1	11^6
1	7	21		35		35		21	7	1	11^7
1	8	28	56		70		56	28	8	1	11^8
1	9	36	84	126		126	84	36	9	1	11^9
1	10	45	120	210	252	210	120	45	10	1	11^{10}

Figure 4.7

The oblique path to the left of the line marked in figure 4.8 indicates the *natural numbers*. Then, to the right of it (and parallel to it), you will notice the *triangular numbers*: 1, 3, 6, 10, 15, 21, 28, 36, 45, ...

From the Pascal triangle (figure 4.8), you should notice how the triangular numbers evolve from the sum of the natural numbers. That is, the sum of the natural number to a certain point may be found by simply looking to the number below and to the right of the last number to be summed (for example, the sum of the natural numbers from 1 to 7 is below and to the right of the 7, which gives a sum of 28.

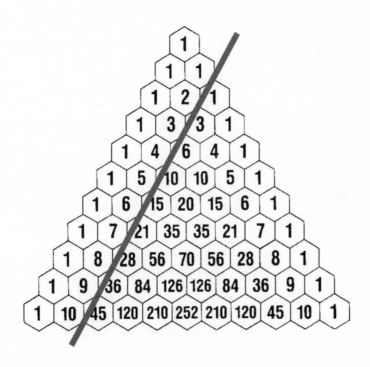

Figure 4.8

The square numbers are embedded as consecutive triangular numbers: $1 + 3 = 4, 3 + 6 = 9, 6 + 10 = 16, 10 + 15 = 25 =, 15 + 21 = 36$, and so on.

We can also find the square numbers in groups of four; we leave these to you to identify, but we offer you one clue here: $1 + 2 + 3 + 3 = 9$, $3 + 3 + 6 + 4 = 16$, $6 + 4 + 10 + 5 = 25$, $10 + 5 + 15 + 6 = 36$, and so on.

In the Pascal triangle shown in figure 4.9, when we add the numbers along the lines indicated, we find that we have located the Fibonacci numbers: 1, 1, 2, 3, 5, 8, 13, 21, 34, 55, 89, 144, . . .

There are many more numbers embedded in the Pascal triangle. You may wish to find the pentagonal numbers: 1, 5, 12, 22, 35, 51, 70, 92, 117, 145, . . . The turf is fertile. Finding more gems in this triangular arrangement of numbers is practically boundless!

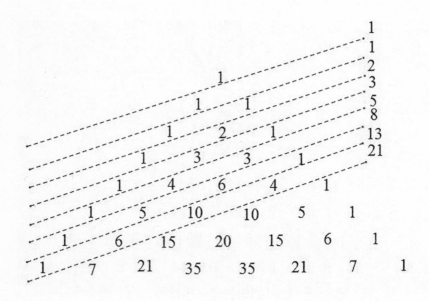

Figure 4.9

RANDOM WALKS

The term "random walk" was first introduced by the English mathematician Karl Pearson (1857–1936). A random walk is a mathematical model used to describe the motion of molecules in a gas, fluctuations of stocks markets, and many other processes in which some apparent or true randomness is involved. The simplest version of a (one-dimensional) random walk can be imagined as a game in which the player (the walker) starts at the initial position 0 and each move requires to take either a step forward (+1) or a step backward (–1). These are the only two options, and the walker must choose at random, for instance, by tossing a coin. If it lands heads up, one step forward has to be made; if it lands tails up, one step backward is in order. Both outcomes occur with equal probabilities. After n moves, the position of the walker will correspond to some integer, which we denote by $X(n)$. In probability theory, this is called a discrete random variable. It is discrete because it can only take on integer values; and it is random because the value of $X(n)$ is subject to variations due to chance (how many times the coin came up heads or tails, respectively).

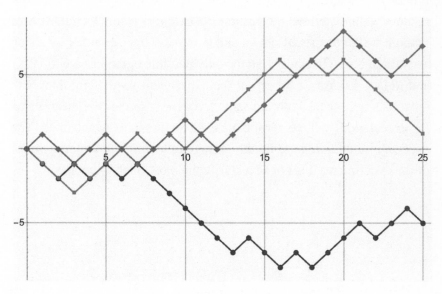

Figure 4.10

Three examples of a one-dimensional random walk as described above are shown in figure 4.10. Here, the horizontal coordinate is the number of steps taken and the vertical coordinate corresponds to $X(n)$, the position of the walker with respect to the starting point. When looking at these examples, we might ask: "How far away from the starting point will the walker be, on average, after n moves?" In probability theory, the average value of a random variable is called its expected value (or expectation) and is usually denoted by E. The expected value of $X(n)$ is written as $E[X(n)]$, and we can think of it as the long-run average, or mean, of $X(n)$, if we repeat an n-step random walk a large number of times. Based on the three 25-step random walks shown in figure 4.10, we can compute the arithmetic mean of $X(25)$ as $\dfrac{7 + 1 - 5}{3} = 1$. What would we get if we performed the same experiment a hundred or a thousand times? Well, since in each move the walker is equally likely to go either forward or backward, we must expect that there will be no progress on the average. In other words, we have $E[X(n)] = 0$ for any n. However, this does not mean that $X(n) = 0$ is the most likely outcome of our random walk experiment. Consider the case $n = 1$, in which the whole random walk consists of only a single move. This can either be a step forward or a step backward, so the only possible outcomes are $X(1) = 1$ and $X(1) = -1$. Yet, if we repeat the experiment very often, then both outcomes will occur with the same frequency, so that the mean value (or expectation) will be zero, that is, $E[X(1)] = 0$, as we claimed. The expectation value represents the average outcome in many repetitions of the experiment; it is not necessarily the most likely outcome.

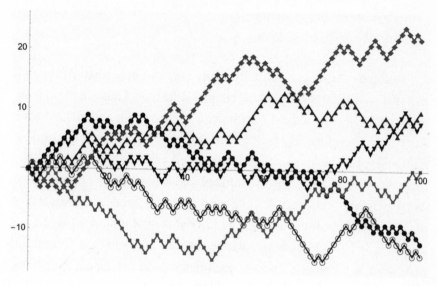

Figure 4.11

Figure 4.11 shows six random walks for $n = 100$ and equal probabilities for forward and backward steps. From the plotted paths we get the impression that the distance to the starting point increases with the number of steps taken. This distance is given by the absolute value $|X(n)|$. Recall that the absolute value $|x|$ of a real number x is defined as the nonnegative value of x without regard to its sign; for example, $|-1| = 1$.

To find out whether $|X(n)|$ really increases with n, we have to calculate its expectation $E\left[\left|X(n)\right|\right]$. For this purpose we use a simple, but important, observation: From position $X(n)$ on, the walker must proceed either to $X(n) + 1$ or to $X(n) - 1$, so the only possible relations between $X(n + 1)$ and $X(n)$ are:

$$X(n + 1) = X(n) + 1,$$
$$X(n + 1) = X(n) - 1.$$

If we square both equations, we obtain:

$$X(n + 1)^2 = X(n)^2 + 2X(n) + 1,$$
$$X(n + 1)^2 = X(n)^2 - 2X(n) + 1.$$

Both possibilities are equally likely and will occur half of the time in a sufficiently large number of independent steps. Thus, on the average we expect $X(n + 1)^2$ to be the arithmetic mean of both variants, which is $X(n)^2 + 1$. Therefore, $E[X(n + 1)^2] = E[X(n)^2 + 1]$, which is the same as $E[X(n)^2] + 1$. This means that the expected value of the square of $X(n)$ increases by 1 in each step. We already know that $X(1) = \pm 1$, and hence, $X(1)^2 = 1$, implying that $E[X(1)^2] = 1$, as well. Since the expectation of $X(n)^2$ increases by 1 in each additional step, we have $E[X(2)^2] = 2$ and $E[X(3)^2] = 3$ and so on, which we can write in one formula as $E[X(n)^2] = n$. However, since we are interested in the distance itself, we have to take the square root to get $E[|X(n)|] = \sqrt{n}$, a simple, but nonetheless surprising result. For an n-step random walk with step size 1, the expected distance from the starting point after n steps is equal to the square root of n. For an arbitrary step size s, the expected distance would be $\sqrt{n}$ times s.

To conclude, the plots in figure 4.11 were not misleading, and the "average random walker" will indeed move away from the origin as the number of steps increases. More precisely, after n steps we expect to find the walker at a distance $\sqrt{n}$ from the origin. This is an important result because it shows that random fluctuations alone are sufficient to significantly increase or decrease stock prices. Moreover, the random movement of particles in a gas or in a fluid can be described as three-dimensional random walks and our result reveals a typical feature of such processes: A tiny particle dropped into a bath tub filled with water will move away from its initial position and on average gain a distance from its starting point that is proportional to the square root of the elapsed time. In other words, the random thermal movement of molecules, called Brownian motion, will eventually lead to an equal distribution of particles dissolved in a solution. Stirring your café au lait is not necessary, if you can wait long enough.

THE POKER WILD-CARD PARADOX

Any poker player knows about the hierarchy of hands: a full house beats a straight, a straight beats three of a kind, three of a kind beats a pair. It seems obvious to anyone with even a bit of knowledge of elementary probability that the hierarchy of hands should be derived from the probability of being dealt such hands from a random shuffle. This is certainly true for a normal game of poker. Surprisingly, it turns out not to be possible to define a hierarchy with this property if wild cards are introduced into the game, as is often the case in friendly games.

As we mentioned earlier, the origins of probability theory derive from the eternal wish to get an edge on your opponent in the context of some gambling pursuit (as was the case when Blaise Pascal and Pierre de Fermat were corresponding on the ideas that would lead to the fundamental concepts of modern probability theory).

Arguably, the most popular modern casino game is poker. Top players are famous from their TV appearances, and at some point almost everyone has tried his or her hand at the game, even if not actually playing for high stakes. For many people there is an undeniable fascination for this game, which perhaps emanates more from the psychological aspects of the betting process than from the game itself.

As was mentioned at the outset, the order of the value of poker hands when the game is played with a standard 52-card deck is determined by the probabilities of the hands. The standard hierarchy, from best to worst, is

– royal flush
    – straight flush
        – four of a kind
            – full house
                – flush
                    – straight
                        – three of a kind
                            – two pair
                                – one pair
                                    – nothing ("junk")

(Note that we assume here that the reader is familiar with these terms. If not, it will not be hard to find someone happy to explain them, or a website with all the definitions. Just don't let yourself get dragged into any of the gambling websites.)

In school, you will likely have learned methods to calculate the probabilities of each of these hands. In order to do this, we must first note that the total number of possible poker hands is equal to the number of ways to choose five cards from a deck of 52. This is usually written as the binomial coefficient $\binom{52}{5}$, and it is equal to 2,598,960, or approximately 2.6 million ways. Since four of these are royal flushes (one in each suit), the probability of a royal flush is about four out of 2.6 million, or 0.000001539.

Calculating the probabilities of the other hands is a bit more involved, but it is certainly the case that the more valuable hands always correspond to the smaller probabilities.

What you probably didn't learn in school is the fascinating fact that if wild cards are introduced to the game, this correspondence between hand values and probabilities does not work; in fact, it cannot work. Many recreational poker players (not the pros, of course) prefer to play with wild cards in the deck, as this gives them the feeling of having a better chance of getting a valuable hand. These wild cards can be the two jokers generally included in a deck of cards (yielding

54 cards to play with, rather than the standard 52), or cards already included. (Some people play with twos wild, for instance.) If you happen to be one of these people, you may want to rethink your stance on the matter after giving some thought to the probabilistic consequences of wild cards.

It is immediately obvious that the probabilities of certain hands change if wild cards are included. Adding two jokers to the 52-card deck, for instance, changes the number of possible hands to $\binom{54}{5}$ (which is equal to 3,162,510), and this certainly has an effect on the probabilities of all possible hands, since they all depend on the total number of hands. Redefining twos as wild does not change the number of possible hands in the sense that there are still $\binom{52}{5}$ ways to choose five cards from the 52-card deck, but holding wild cards means that their values can be defined by the player, altering the number of possible hands of each type.

Furthermore, the inclusion of wild cards means that a new high-value hand is introduced that is not possible under normal circumstances, namely five of a kind. This is generally considered the highest valued hand, since it is even rarer than a royal flush. Since there is now one more type of hand to consider, the relative probabilities of each of the possible hands must also be influenced by this new hand.

The amazing thing about introducing wild cards of any type lies in the fact that it is not possible to define a hierarchy of hands that is consistent with the resulting probabilities. No matter how the values of the hands are defined in this case, there will always be a hand A ranked higher than a hand B, such that the probability of hand A is simultaneously higher than that of hand B.

This is a consequence of the fact that the higher ranking of a hand will necessarily lead to a player defining a wild card in such a way that the higher-ranked hand results. This, in turn, makes that hand more likely to be dealt, increasing its probability. To see how this can happen, consider the following examples.

Take, for instance the hand

5♥   6♦   7♥   joker   joker.

Depending on the player's choice, this could be played as a straight with 9 high (if the jokers are defined as a 9 and an 8), or a straight with 7 or 8 high (with the jokers as 3 and 4, or 4 and 8 respectively), or as three of a kind (three 7s or three 6s or three 5s), two pair or one pair in many ways, or even as junk (taking one joker as a 9 and one as a 10, for instance).

Holding two wild cards, we have a lot of leeway, but even with just a single one, we can see the implications. Take the hand

5♥   6♦   7♥   7♠   joker.

The joker can be defined as another 7, giving a hand of three of a kind, or as a 5 or 6, giving a hand of two pair. (We ignore the options that define the joker as something giving no advantage.) So, where is the conundrum?

Let us first assume that three of a kind is ranked higher than two pair, as is normal for a standard deck with no wild cards. In this case, the player will choose the more valuable option and call three of a kind. In this case, hands such as this one will always be defined as three of a kind, raising the number of combinations of five cards among the 3,162,510 possible ones that are "worth" three of a kind. As it turns out, the actual calculation is rather lengthy,[2] but can certainly be attempted by any interested reader with a bit of knowledge concerning binomial coefficients. From this point of view, we find that there will actually be more three-of-a-kind hands (over 230,000) than two-pair hands (less than 130,000), which contradicts our basic notion that the more-valuable hand should be the rarer one.

On the other hand, let us assume that we redefine the hierarchy of hands and say that two pairs is worth more than three of a kind. In this case (with the above selection of cards), any player will define the joker as a 6 in order to hold a pair of 7s and a pair of 6s. And, wouldn't you know it, this definition once again yields just what we don't want. This ordering means that there are more hands "worth" two pair (over 300,000) than there are of those "worth" three of a kind (less than 60,000), and we are once again faced with the situation that the higher-ranked hand (two pair in this case) is more common than the lower-ranked one (three of a kind).

In fact, it can be shown that any inclusion of a wild card in the poker deck opens us up to a host of such inconsistencies.[3] This is the Wild-Card Paradox. It makes you want to reconsider ever gambling with wild cards, doesn't it?

CHAPTER 5

COMMON SENSE FROM A MATHEMATICAL PERSPECTIVE

This chapter is intended to inform and to entertain you with topics that are typically not part of the curriculum but perhaps ought to be and could be accommodated in a variety of places. Here, we will show you how some harmless, yet very powerful, problem-solving techniques that can enhance your thinking in mathematics and beyond. We will begin by mentioning some historical facts about the symbols we use in mathematics that seem to be taken for granted, and might help make understanding mathematics more meaningful for you. Beyond the symbols themselves, there are concepts that are often taken for granted as well, such as the concept of infinity, which is oftentimes felt to be too theoretical and difficult to comprehend at the high school level and therefore is not properly introduced. We also include some practical applications of mathematics that could clearly be a part of everybody's thinking. Naturally, we can't cover everything here, but we surely hope to widen your perspective for future endeavors in mathematics.

THE ORIGINS OF SOME MATHEMATICS SYMBOLS

Unfortunately, our schooling introduces mathematical symbols without telling us from where they emanate. Take, for example, the square root symbol ($\sqrt{}$), which is used throughout our study of elementary mathematics. Teachers don't usually explain how we got to this strange-looking symbol. Well, this symbol was first used by the German mathematician Christoff Rudolff (1499–1545) in his book *Coss*, which is an arithmetic book published in Strassburg in 1525. It is speculated that he came upon this symbol from the letter *r* in script, which might have come from the word *radix*, meaning "root."

The first known appearance of the plus (+) and minus (–) signs was in a book by the German mathematician Johannes Widmann (1460–1498) titled *Behende und hüpsche Rechenung auff allen Kauffmanschafft*, which was published in Leipzig in 1489. Here it was not used for addition and subtraction, though, but rather for surpluses and deficits in business problems. There is some controversy as to who the first person was to use the + and the – symbols to represent addition and subtraction. Some say it was Giel Van Der Hoecke in his book titled *Een sonderlinghe boeck in dye edel conste Arithmetica*, which was published in Antwerp in 1514; others say it was the German mathematician Henricus Grammateus, also known as Heinrich Schreyber (1495–1526) in his book titled *Ayn new Kunstlich Buech*, which was published in 1518. The Welsh mathematician Robert Recorde (1512–1558) was the first to use these symbols in the English language in 1557, in his book *The Whetstone of Witte*, in which he wrote, "There be other 2 signes in often use of which the first is made thus + and betokeneth more: the other is thus made – and betokeneth lesse."

We credit the English mathematician William Oughtred (1574–1660) as the first to use the × to represent multiplication, in his book *Clavis Mathematicae* (The Key to Mathematics), written around 1628 and published in London in 1631. The dot (·) was favored to indicate multiplication by the German mathematician Gottfried Wilhelm Leibniz (1646–1716). On July 29, 1698, he wrote in a letter to the Swiss mathematician Johann Bernoulli (1667–1748): "I do not like × as a symbol for multiplication, as it is easily confounded with x; . . . often I simply relate two quantities by an interposed dot and indicate multiplication by $ZC \cdot LM$. Hence, in designating ratio, I use not one point but two points (the colon), which I use at the same time for division."[1]

Despite Leibniz's suggestion to use the colon for division, American books use the obelus ($\div$) to indicate division. This was first used by Swiss mathematician Johann Rahn (1622–1676), in his book *Teutsche Algebra*, which was published in 1659.

It was the Welsh mathematician Robert Recorde, who in 1557 introduced the use of the = sign to avoid writing "is equal to" each time. (See figure 5.1 for the section in which Recorde introduced the equals sign.) And while on the topic of equality, we credit the British mathematician Thomas Harriot (1560–1621) with first using the inequality symbols > and < in a book published posthumously in 1631.

Howbeit, for eafie alteratiõ of *equations.*J will propounde a fewe eraples, bicaufe the ertraction of their rootes, maie the more aptly bee wroughte. And to auoide the tedioufe repetition of thefe woordes : is equalle to : J will fette as J doe often in woorke bfe, a paire of paralleles, or Gemowe lines of one lengthe, thus: ========, bicaufe noe. 2. thynges, can be moare equalle. And now marke thefe nombers.

Figure 5.1

Here you now have the origin of the basic symbols that we always simply take for granted without concern from where they came. Let us now embark on a few thinking exercises oftentimes known as problem-solving strategies.

THE COUNTERINTUITIVE

In the course of mathematics instruction there are times when material is presented that is completely different from what was expected. Too often there is not enough attention paid to these unusual situations that seem to throw off our logical assessments. Yet by paying attention to these counterintuitive situations we can become better observers in everyday life and deal with unusual problem situations more objectively. This sort of situation seems to manifest itself best in the realm of problem solving—in this case, mathematical problem solving.

Let's consider one such situation. Suppose you are presented with a collection of toothpicks arranged as shown in figure 5.2, in which each of the two top and bottom rows and two side columns contains eleven toothpicks.

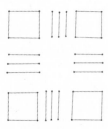

Figure 5.2

We are now asked to remove one toothpick from each row and column and still remain with eleven toothpicks in each row and column. This seems to be impossible, since we are actually removing toothpicks, and yet we are asked to keep the same number of toothpicks in each row and column, as before. As a first attempt, we could remove four toothpicks to get the situation shown in figure 5.3.

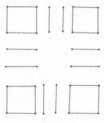

Figure 5.3

But we see that this attempt has failed. Faced with this, we then ask ourselves, how can this possibly be done? Up until now we have dealt with something that is clearly counterintuitive. If this can be done as laid out in figure 5.3, we would have to count some toothpicks twice. In figure 5.4, we see that we've taken a toothpick from the center portion of each of the rows and each of the columns, and we have placed these toothpicks in the corner positions so that they can be counted more than once.

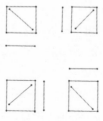

Figure 5.4

Thus, we have achieved our goal of having eleven toothpicks in each of the two outer rows and each of the two outer columns. The notion of things being counterintuitive can be seen in a number of other units throughout the book, such as "The False Positive Paradox," "The Birthday Phenomenon," or "The Monty Hall Problem," and others. This is a topic that merits everyone's attention so that we go through life by analyzing situations in a more critical fashion.

A SURPRISING SOLUTION

One of the important aspects in the teaching of mathematics, particularly at the secondary school level, is problem solving. Unfortunately, too often the procedure of solving mathematical problems is guided as an application of the topic being taught, rather than considering the problem-solving strategy to be used. There are some problems that can be made particularly simple by looking at the problem from a point of view other than the one that the problem seems to guide the reader to consider. We present a problem here, which is one that can be easily solved by merely following the "path" of the problem as it is presented. This is something that most folks would do, as it is quite natural. However, the problem does present a great opportunity to appreciate the point that we would like to make here, that is, to demonstrate how to look at a problem from a different point of view—something that is all too often passed over in the usual school mathematics experience. You might wish to try the problem yourself (without reading ahead) and see whether you fall into the "majority-solvers" group. The solution offered later will probably enchant most readers—as well as provide some valuable future guidance.

The problem: A single elimination basketball tournament (one loss, and the team is eliminated) consists of twenty-five teams competing. How many games must be played until there is a single tournament champion?

Typically the majority-solvers will begin to simulate the tournament by taking two groups of twelve teams playing each other during the first round, with one team drawing a bye (*12 games played*). After the first round, twelve teams will have been eliminated, leaving twelve teams and the one team that drew a bye remaining in the tournament. In the next round, of these thirteen teams, six teams will play another six teams, leaving six winners and the one team that drew a bye (*6 games played*). In round three, of the seven teams remaining, three will play another three, leaving three winners, plus the team that drew a bye

(*3 games played*). In round four, the four remaining teams will play each other (*2 games played*), leaving two remaining teams that will play each other for the championship (*1 game played*). When we now count the number of games played, $12 + 6 + 3 + 2 + 1 = 24$, we find the necessary number of games played to get a champion team. This is a perfectly legitimate problem-solving technique, but it is clearly not the most elegant or efficient method. Unfortunately, we are often not given the opportunity to investigate the solution to such a straightforward problem with an alternative—and perhaps often more elegant—"off the beaten path"—problem-solving technique. Let us now consider just such an alternative solution to the original problem.

A much simpler way to solve this problem, one that most people do not naturally come up with as a first attempt, is to focus only on the losers, and not on the winners, as we have done above. We ask the key question: "How many losers must there be in the tournament with twenty-five teams in order for there to be one winner"? The answer is simple: twenty-four losers. How many games must be played to get twenty-four losers? Naturally, twenty-four games. So there you have the answer, very simply done.

Now most people will ask themselves, "Why didn't I think of that?" The answer is, it was contrary to the type of training and experience we have had. Becoming aware of the strategy of looking at the problem from a different point of view may sometimes reap nice benefits, as was the case here. You may never know which strategy will work; so just try one and see. In this case, we simply looked at the opposite of what everyone else focuses on—we considered the losers instead of the winners. Doing so gave us a handy way to grasp the solution.

There is an interesting alternative to the above solution that can be put as follows: supposing of the twenty-five teams, there is clearly one that is vastly superior to all the others, such as one of the NBA teams. We could have each of the remaining twenty-four teams playing this superior team and, of course, they will lose the game. Here again, you can see that only twenty-four games are required to get the champion, in this case, superior team.

DON'T "WINE" OVER THIS PROBLEM—
A PROBLEM-SOLVING APPROACH

In school, there is a great deal of emphasis put on solving mathematical problems. Typically, problem-solving experiences that one faces at the secondary school level are such that the problems are made to fit into certain categories, and students are taught to deal with these types of problems in a somewhat mechanical fashion. Unfortunately, this does not do much for training students to solve problems in a real-world situation. To show a problem and its solution that should have been presented in the schools, and typically wasn't, provides us the opportunity to make up for past omissions. The problem reads very simply and yet can be confusing to some. However, it is the elegance of the solution that needs to be stressed and appreciated. Here's the problem:

We have two one-gallon bottles. One contains a quart of red wine and the other, a quart of white wine. We take a tablespoonful of red wine and pour it into the white wine. Then we take a tablespoon of this new mixture (white wine and red wine) and pour it into the bottle of red wine. Is there more red wine in the white wine bottle, or more white wine in the red wine bottle?

To solve the problem, we can figure this out in any of the usual ways—often referred to in the high school context as "mixture problems"—or we can use some clever logical reasoning and look at the problem's solution as follows:

The simplest solution to understand and the one that demonstrates a very powerful problem-solving strategy is that of *using extremes*. We use this kind of reasoning in everyday life when we resort to the option: "such and such would occur in a worst-case scenario, so we can decide to . . ."

Let us now employ this strategy for the above problem. To do this, we will consider the tablespoonful quantity to be a bit larger. Clearly the outcome of this problem is independent of the quantity transported. So we will use an *extremely* large quantity. We'll let this quantity actually be the *entire* one quart. That is, following the instructions given the problem statement, we will take the entire amount (one quart of red wine), and pour it into the white-wine bottle. This mixture is now 50 percent white wine and 50 percent red wine. We then pour one quart of this mixture back into the red-wine bottle. The mixture is now the same in both bottles. Therefore, there is as much white wine in the red wine bottle as there is red wine in the white wine bottle!

We can consider another form of an extreme case, where the spoon doing the wine transporting has a zero quantity. In this case, the conclusion follows immediately: There is as much red wine in the white wine bottle as there is white wine in the red wine bottle, that is, zero!

This solution method can be very significant in the way you might approach future mathematics problems and even how you may analyze future everyday decision making.

ORGANIZED THINKING

Oftentimes we are faced with a problem that at first sight seems a bit overwhelming. Unfortunately, such problems don't often appear as part of the typical school mathematics curriculum. However, problems of this sort can be helpful in approaching everyday life issues with which we may be confronted, and at first sight don't lend themselves easily to a logical or simple solution strategy. Let's consider a few of these now.

Asked how many different triangles appear in the diagram shown in figure 5.5, we tend to begin to count and then quickly find ourselves confused as to whether or not we have already counted a particular triangle.

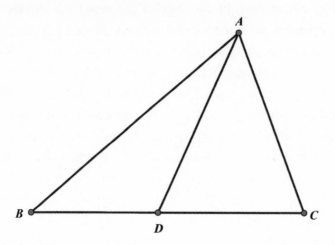

Figure 5.5

Therefore, it helps to reconstruct the diagram step by step, each time adding another line and counting those triangles that depend on the newly added line for a side. We shall begin with just one line in the triangle *ABC*, as shown in figure 5.5.

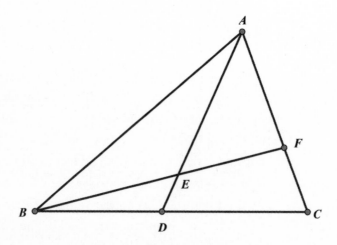

Figure 5.6

In figure 5.5 we find that there are only three triangles, namely, $\triangle ABC$, $\triangle ABD$, and $\triangle ACD$. We now add another line, BF, and count the number of triangles in figure 5.6.

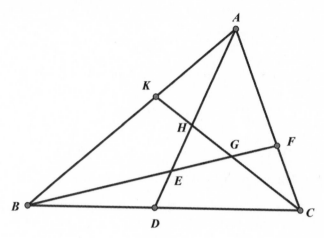

Figure 5.7

With the segment BF, we find that the following triangles use that segment or part of it for a side: $\triangle BED$, $\triangle ABE$, $\triangle ABF$, $\triangle AEF$, and $\triangle BFC$. We now add the third segment to complete the figure as it was originally presented (refer back to figure 5.5), and list the triangles that depend on segment CK for a side or part of the segment for a side. These are triangles: $\triangle BGC$, $\triangle BKC$, $\triangle HEG$, $\triangle DHC$, $\triangle BKG$, $\triangle AKH$, $\triangle AHC$, $\triangle GFC$, and $\triangle AKC$ (see fig. 5.7).

In all, we now have seventeen triangles in the original figure, the counting of which was made simpler by reconstructing the figure piece by piece, and each time counting the new triangles that appeared with the additional line segments. This is a useful strategy, in general, when the problem might appear overwhelming at first sight.

Another problem of this sort, that is, a problem that can be most easily solved by approaching it in a piecemeal fashion, is the following:

If, on the average, a hen and a half can lay an egg and a half in a day and a half, how many eggs should six hens lay in eight days?

To approach this problem in a logical and progressive fashion, we will double only two of the three variables of the time, as follows:

We begin with what was presented in the original problem:

$\frac{3}{2}$ hens lay $\frac{3}{2}$ eggs in $\frac{3}{2}$ days

By doubling the number of hens (leaving the number of days constant):

3 hens lay 3 eggs in $\frac{3}{2}$ days

By doubling the days (leaving the number of hens constant):

3 hens lay 6 eggs in 3 days

Taking one third of the days (leaving the number of hens constant):

3 hens lay 2 eggs in 1 day

Doubling the number of hens (leaving the number of days constant):

6 hens lay 4 eggs in 1 day

Therefore, in eight days, by multiplying the number of days by 8 (leaving the number of hens constant), we get our sought-after result:

6 hens lay 32 eggs in 8 days.

Notice that we never dealt with more than two of the variables at any one step in this process. This simplifies a rather confusing problem.

The logical thinking that these two problems requires is often neglected from the school curriculum, but it has a true value when dealing with everyday problems.

SUCCESSIVE PERCENTAGES

Most folks find percentage problems to have long been a nemesis, and unfortunately percentages are often tediously taught in the schools. Problems get particularly unpleasant when multiple percentages need to be processed in the same problem. It is our hope that this section will turn this onetime nemesis into a delightfully simple arithmetic algorithm that affords lots of useful applications and provides new insights into successive percentage problems. This not very well-known procedure could well enchant you. Let's begin by considering the following problem:

Wanting to buy a coat, Charles is faced with a dilemma. Two competing stores next to each other carry the same brand of coat with the same list price, but with two different discount offers. Store A offers a 10 percent discount year round on all its goods, but on this particular day offers an additional 20 percent on top of their already-discounted price. Store B simply offers a discount of 30 percent on that day in order to stay competitive. How do the two discounted prices compare?

At first glance, you may assume there is no difference in price, since $10 + 20 = 30$, which would lead you to believe that the two stores are offering the same discount. Yet with a little more thought you may realize that this is not correct, since in store A only 10 percent is calculated on the original list price; the 20 percent is then calculated on the lower price. In contrast, at store B, the entire 30 percent is calculated on the original price. Now the question to be answered is, what is the difference, if any, is there between the discounts in store A and store B?

One expected procedure might be to assume the cost of the coat to be $100, calculate the 10 percent discount, yielding a $90 price, and an additional 20 percent off the $90 price (or $18) will bring the price down to $72. In store B, the 30 percent discount on $100 would bring the price down to $70, giving a discount difference of $2, which in this case is 2 percent. This procedure, although correct and not too difficult, is a bit cumbersome and does not always allow full insight into the situation.

An interesting and quite unusual procedure is provided here, and such a procedure could have been presented at school as well. Let us now follow this method for obtaining a single percentage discount (or increase) equivalent to two (or more) successive discounts (or increases).

1. Change each of the percentages involved into decimal form:

   .20 and .10

2. Subtract each of these decimals from 1.00:

   .80 and .90 (for an increase, add to 1.00)

3. Multiply these differences:

   $(.80)(.90) = .72$

4. Subtract this number (i.e., .72) from 1.00:

   $1.00 - .72 = .28$, which represents the combined *discount*

(If the result of step 3 is greater than 1.00, subtract 1.00 from it to obtain the percentage of *increase*.)

When we convert .28 back to percent form, we obtain 28 percent, the equivalent of successive discounts of 20 percent and 10 percent. This combined percentage, 28 percent, differs from 30 percent by 2 percent.

Following the same procedure, you can also combine more than two successive discounts. In addition, successive increases, combined or not combined with a discount, can also be accommodated with this procedure by adding the decimal equivalent of the increase to 1.00, whereas the discount was subtracted from 1.00 and then continued with the procedure in the same way. If the end result turns out to be greater than 1.00, then this end result reflects an overall increase rather than the discount that was found in the above problem.

This procedure not only streamlines a typically cumbersome situation but also provides some insight into the overall picture. Consider, for example, the question "Is it advantageous to the buyer in the above problem to receive a 20 percent discount and then a 10 percent discount, or to receive the reverse, a 10 percent discount and then a 20 percent discount?" The answer to this question is not immediately intuitively obvious. Yet, since the procedure just presented shows that the calculation is merely multiplication, which is a commutative operation, we can see that there is no difference between the two options. So here you have a delightful algorithm for combining successive discounts or increases or combinations of these. Not only is it useful, but it also gives you some newfound power in dealing with percentages.

THE RULE OF 72

The school curriculum at some point has students compute compound-interest problems with a formula such as $A = P\left(1 + \dfrac{r}{100}\right)^n$, where A is the resulting amount of money, P is the principal invested for n interest periods at r percent annually. What teachers neglect to mention is that there is a curious little scheme that works well and is somewhat puzzling to verify. It is called the "Rule of 72," and it may still generate some interest in its simplicity.

The "Rule of 72" states that, roughly speaking, money will double in $\dfrac{72}{r}$ years when it is invested at an annual compounded interest rate of r percent. So, for example, if we invest money at an 8 percent compounded annual interest rate, it will double its value in $\dfrac{72}{8} = 9$ years. Similarly, if we leave our money in the bank at a compounded rate of 6 percent, it would take $\dfrac{72}{6} = 12$ years for this sum to double its value. As we said, the beauty of this technique lies in its simplicity. For the interested reader we provide some explanation of why this procedure actually works.

We begin by considering once again the compound-interest formula: $A = P\left(1+\dfrac{r}{100}\right)^n$.

We need to investigate what happens when $A = 2P$.

The above equation then becomes:

$$2P = P\left(1+\frac{r}{100}\right)^n, \text{ or } 2 = \left(1+\frac{r}{100}\right)^n \qquad \text{(I)}$$

It then follows that $n = \dfrac{\log 2}{\log\left(1+\dfrac{r}{100}\right)}$. $\qquad$ (II)

With the help of a scientific calculator, let us make a table of values from the above equation:

r	n	nr
1	69.66071689	69.66071689
3	23.44977225	70.34931675
5	14.20669908	71.03349541
7	10.24476835	71.71337846
9	8.043231727	72.38908554
11	6.641884618	73.0607308
13	5.671417169	73.72842319
15	4.959484455	74.39226682

If we take the arithmetic mean (the average) of the *nr* values, we get 72.04092314, which is quite close to 72, and so our "Rule of 72" seems to be a very close estimate for doubling money at an annual interest rate of *r* percent for *n* interest periods.

An ambitious reader might try to determine a "rule" for tripling and quadrupling money, similar to the way we dealt with the doubling of money. The analogous equation for *k*-tupling would be

$$n = \frac{\log k}{\log\left(1+\dfrac{r}{100}\right)},$$

which for $r = 8$ gives the value for $n = 29.91884022$ (log k).

Thus, $nr = 239.3507218$ log k, which for $k = 3$ (the tripling effect) gives us $nr = 114.1993167$. We could then say that for tripling money we would have a "Rule of 114."

However far this topic is explored, the important issue here is that the common "Rule of 72" can nicely complement your knowledge of school mathematics.

A MATHEMATICAL CONJECTURE

Much of the mathematics that is taught in school is somehow justified with some kind of logical argument or proof. There are a number of phenomena in mathematics that appear to be true but have never been properly substantiated, or proved. These are referred to as mathematical conjectures. In some cases, computers have been able to generate enormous numbers of examples to support a statement's veracity, but that (amazingly) does not make it true for all cases. To conclude the truth of a statement, we must have a legitimate proof that it is true for *all* cases!

Perhaps one of the most famous mathematical conjectures that has frustrated mathematicians for centuries was presented by the German mathematician Christian Goldbach (1690–1764) to the famous Swiss mathematician Leonhard Euler (1707–1783) in a June 7, 1742, letter, in which he posed the following statement, which to this day has yet to be proved. Commonly known as *Goldbach's conjecture*, it states the following: Every even number greater than 2 can be expressed as the sum of two prime numbers.

We can begin with the following list of even numbers and their prime-number sums and then continue it to convince ourselves that it continues on—apparently—indefinitely.

Even Numbers Greater Than 2	Sum of Two Prime Numbers
4	2 + 2
6	3 + 3
8	3 + 5
10	3 + 7
12	5 + 7
14	7 + 7
16	5 + 11
18	7 + 11
20	7 + 13
. . .	. . .
48	19 + 29
. . .	. . .
100	3 + 97

There have been substantial attempts by famous mathematicians to try to prove or justify this conjecture. In 1855, A. Desboves verified Goldbach's conjecture for all even numbers up to 10,000. Yet, in 1894, the famous German mathematician Georg Cantor (1845–1918) (regressing a bit) showed that the conjecture was true for all even numbers up to 1,000. It was then shown by N. Pipping in 1940 to be true for all even numbers up to 100,000. By 1964, with the aid of a computer, it was justified up to 33,000,000, and in 1965 this was extended to 100,000,000. Later, in 1980, the conjecture was justified up to 200,000,000. Then, in 1998, the German mathematician Jörg Richstein showed that Goldbach's conjecture was true for all even numbers up to 400 trillion. As of May 26, 2013, Tomás Oliveira e Silva justified the conjecture for numbers up to $4 \cdot (10^{17})$. Prize money of $1,000,000 has been offered for a proof of this conjecture. To date, this has not been claimed, as the conjecture has never been proved for *all* cases.

Goldbach also had a second conjecture, which was proved true by the Peruvian mathematician Harald Helfgott in 2013 and is stated as follows: Every odd number greater than 5 can be expressed as the sum of three primes.

Again, we shall present a few examples and let you continue the list as far as you wish.

Odd Numbers Greater Than 5	Sum of Three Prime Numbers
7	2 + 2 + 3
9	3 + 3 + 3
11	3 + 3 + 5
13	3 + 5 + 5
15	5 + 5 + 5
17	5 + 5 + 7
19	5 + 7 + 7
21	7 + 7 + 7
. . .	. . .
51	3 + 17 + 31
. . .	. . .
77	5 + 5 + 67
. . .	. . .
101	5 + 7 + 89

Of course, if the first conjecture is true, this one must be as well, since subtracting the prime number 3 from an odd number yields an even number. If this is expressible as the sum of two primes, the original number can certainly be expressed as the sum of three primes.

These unsolved problems have tantalized many mathematicians over the centuries, and although only the latter one has been proved true and no counterexample has been found for the first one, even with the help of computers, there is a strong suggestion that even the first conjecture would be true. Interestingly, the efforts to prove them have led to some significant discoveries in mathematics that might have gone hidden without this impetus. They provoke us as well as provide sources of entertainment.

UNEXPECTED PATTERNS

The sequences and series that most are exposed to in school are the ones that are usually recognized by a common pattern. For example, when presented with the beginning five numbers of the following sequence, most people would guess that 32 is the next number. Yes, that would be fine. However, when the next number is given as 31 (instead of the expected 32), cries of "wrong!" are usually heard. Amazingly, this is also a correct answer, and **1, 2, 4, 8, 16, 31, ...** can be a legitimate sequence. This is an aspect of mathematics to which we should all be exposed: our intuition is not the only criterion for determining the correctness of the situation. In mathematics, we can prove things to be true that seem to be counterintuitive.

The task now is to justify the legitimacy of this sequence. It would be nice if it could be done geometrically, as that would give convincing evidence of a physical nature. We will do that later, but in the meantime, let us first find the succeeding numbers in this "curious" sequence.

We shall set up a chart showing the differences between terms of a sequence, beginning with the given sequence up to 31, and then work backward once a pattern is established, which we find at the third differences.

	c1	c2	c3	c4	c5	c6	c7	c8	c9	c10	c11
Original Sequence	1		2		4		8		16		31
First Difference		1		2		4		8		15	
Second Difference			1		2		4		7		
Third Difference				1		2		3			
Fourth difference					1		1				

With the fourth differences forming a sequence of constants, we can reverse the process by turning the table upside down and extend the third differences a few more steps with 4 and 5.

	c1	c2	c3	c4	c5	c6	c7	c8	c9	c10	c11	c12	c13	c14	c15
Fourth Difference					1		1		1		1				
Third Difference				1		2		3		4		5			
Second Difference			1		2		4		7		11		16		
First Difference		1		2		4		8		15		26		42	
Original Sequence	1		2		4		8		16		31		57		99

The bold-type numbers are those that were obtained by working backward from the third difference sequence. We can then see that the next numbers of the given sequence are 57 and 99. The general term of this sequence is a fourth-power expression, since we had to go to the fourth differences to get a constant. The general term (the nth term) is:

$$\frac{n^4 - 6n^3 + 23n^2 - 18n + 24}{24}.$$

You might get the impression from this little exercise above that this sequence is a contrived arrangement and not one that has some mathematical significance. To dispel this false concept, consider the Pascal triangle shown in figure 5.7.[2]

```
                    1
                   1 1
                  1 2 1
                 1 3 3 1
                1 4 6 4 1
              1 5 10 10 5 1
            1 6 15 20 15 6 1
          1 7 21 35 35 21 7 1
        1 8 28 56 70 56 28 8 1
```

Figure 5.8

Consider the horizontal sums of the rows of the Pascal triangle to the right of the bold line shown in figure 5.7. Curiously enough, these sums, 1, 2, 4, 8, 16, 31, 57, 99, 163, result in our newly developed sequence.

A geometric interpretation should further help to convince you of the beauty and consistency inherent in mathematics. To do this we shall make a chart of the number of regions into which a circle can be partitioned by joining points on the circle. You might want to try this by actually partitioning a circle. Just make sure that no three lines meet at one point, or else you will lose a region.

Number of Points on the Circle	Number of Regions into Which the Circle Is Partitioned
1	1
2	2
3	4
4	8
5	16
6	31
7	57
8	99

Now that you can see that this unusual sequence seems well embedded in various other mathematical aspects, we have provided an aspect of the subject that may have been missed during school instruction of mathematics.

AN INFINITY CONUNDRUM

We all learned the commutative property for addition, namely $1 + 2 = 2 + 1$. However, when we look at an infinite series such as the following; $1 - 1 + 1 - 1 + 1 - 1 + 1 - 1 + \ldots$, we can find the sum by pairing the numbers as

$$(1 - 1) + (1 - 1) + (1 - 1) + (1 - 1) + \ldots = 0 + 0 + 0 + 0 + 0 + \ldots = 0.$$

While on the other hand, we could have also paired the numbers as follows:

$$1 + (-1 + 1) + (-1 + 1) + (-1 + 1) + (-1 + 1) + \ldots = 1 + 0 + 0 + 0 + 0 + \ldots = 1.$$

Here we have the same series of numbers resulting with two different sums, depending on how we can pair the consecutive members of the series. Recall that we are merely using the associative property, which we know holds true in our number system. This conundrum enticed the Italian mathematician Luigi Guido Grandi (1671–1742) to show that he could compromise between the two values, 1 and 0, and get their average, $\frac{1}{2}$, as the sum of the series in the following way:

Let $S = 1 - 1 + 1 - 1 + 1 - 1 + 1 - 1 + \ldots$. We can then consider that since the series goes infinitely far, we can write the series as follows: $S = 1 - (1 - 1 + 1 - 1 + 1 - 1 + 1 - 1 + \ldots)$. We then notice that the value inside the parentheses is also equal to S.

This allows us to write the value of $S = 1 - S$, which in turn is $2S = 1$, or $S = \frac{1}{2}$.

Let's now look at another way to consider this infinite series by considering partial sums as follows:

$$S_1 = 1$$
$$S_2 = 1 - 1 = 0$$
$$S_3 = 1 - 1 + 1 = 1$$
$$S_4 = 1 - 1 + 1 - 1 = 0$$
$$S_5 = 1 - 1 + 1 - 1 + 1 = 1$$

Since the partial sums fluctuate between 1 and 0, it would appear that the series will not converge to a specific value, even at infinity. This would leave us with a meaningless infinite series.

Suppose we now consider the following series, which is called an *alternating harmonic series*:

$$H = 1 - \frac{1}{2} + \frac{1}{3} - \frac{1}{4} + \frac{1}{5} - \frac{1}{6} + \frac{1}{7} - \frac{1}{8} + \ldots$$

If we now take partial sums as we did before, you will recognize an interesting occurrence.

$$S_1 = 1$$

$$S_2 = 1 - \frac{1}{2} = .5000$$

$$S_3 = 1 - \frac{1}{2} + \frac{1}{3} = .8333\ldots$$

$$S_4 = 1 - \frac{1}{2} + \frac{1}{3} - \frac{1}{4} = .5833\ldots$$

$$S_5 = 1 - \frac{1}{2} + \frac{1}{3} - \frac{1}{4} + \frac{1}{5} = .7833\ldots$$

As the number of the terms in the sequence increases, the sum begins to get closer and closer to .693147 . . . , which is the natural log of 2 (written as *ln*2). However, once again we find that by adding the numbers in this alternating harmonic sequence, we would get a variety of other numbers. For example, suppose we group the terms of the sequence as follows:

$$H = \left(1 - \frac{1}{2} - \frac{1}{4}\right) + \left(\frac{1}{3} - \frac{1}{6} - \frac{1}{8}\right) + \left(\frac{1}{5} - \frac{1}{10} - \frac{1}{12}\right) + \cdots$$

Simplifying in each of the parentheses, we get:

$$H = \left(\frac{1}{2} - \frac{1}{4}\right) + \left(\frac{1}{6} - \frac{1}{8}\right) + \left(\frac{1}{10} - \frac{1}{12}\right) + \cdots$$

We can then factor out $\frac{1}{2}$ from each of the parentheses to get the following:

$$H = \frac{1}{2}\left(1 - \frac{1}{2} + \frac{1}{3} - \frac{1}{4} + \frac{1}{5} - \frac{1}{6} + \frac{1}{7} - \frac{1}{8} + \cdots\right),$$

and we notice the reappearance of the alternating harmonic series within the parentheses. Yet this time the same sum (H) is half that of the previously calculated value. That is, $H = \frac{1}{2}H$. This should be motivation for further study. These are just some of the conundrums we have to face when dealing with infinity. It is a concept that should not be taken lightly, and yet it gives us greater insight into mathematics beyond what we are shown in the school curriculum.

THE CONCEPT OF INFINITY

The concept of infinity is something that teachers often choose to avoid, because it is very difficult to comprehend—especially for a young mind that has not yet fully matured. It is difficult to understand how the set of all natural numbers—an infinite set—can have the same size as the set of positive even numbers. After all, the positive odd numbers are missing. However, it can be argued that for every natural number n there is a positive even number $2n$. Thus, the two infinite sets can be said to have the same number of elements, since we can set up a one-to-one correspondence between the members of the two sets of numbers. In retrospect, could such an argument be accepted at the school level by the majority of the population? It is, after all, rather counterintuitive to consider two sets to be equal in size when one clearly contains the other. Furthermore, it is often said that if a monkey were to sit at a keyboard and randomly hit individual keys for an *infinite* duration of time, he would produce all of Shakespeare's works in the order in which he wrote them and much more. That just further complicates our understanding of the concept of infinity.

Then there is the famous paradox of the Grand Hotel originally devised by the German mathematician David Hilbert (1862–1943). This theoretical hotel would have an infinite number of rooms lined up along a corridor: Room 1, Room 2, Room 3, and so forth, continuing without end. Assume that one night all the rooms are occupied and a potential guest arrives in need of a room. The receptionist would actually be able to find accommodations for this new guest. To do that, he would move the occupant from Room 1 to Room 2, and the occupant from Room 2 to Room 3, and the occupant from Room 3 to Room 4, and so forth, for all the guests in the infinite corridor. This paradox can be further extended to a situation in which a bus load of potential hotel guests arrives at a hotel that has infinitely many rooms, all of which are occupied. Here again these guests can be accommodated in a similar manner because of infinity's peculiarity. This can be taken to a further level in which an infinite number of buses arrive, each of which has an infinite number of guests to be accommodated. Even this situation can be handled because of the peculiarity of the concept of infinity.

There are many paradoxes that can be constructed around the concept of infinity. There is the famous paradox by Zeno (490–425 BCE), which, in simple form, would indicate that someone walking toward a wall, and each time going half the distance to the wall, would actually never reach the wall, because he would have to go halfway first for an infinite number of halfway points. While we are speaking of infinity, it is interesting to note that the symbol of infinity (∞) was first introduced in 1655 by the English mathematician John Wallis (1616–1703), and it is still universally used today.

When asked how you can create a set larger than the infinite set of natural numbers one possible construct is to consider the set of all subsets of this infinite set, and that would be a larger set. There are countless examples that can be shown to a school audience regarding the concept of infinity; however, the question is, To what extent can a not yet fully mature mind genuinely understand the complexities related to this concept? Let's explore one example of how the concept of infinity sometimes leads us counterintuitive situations Take, for example, the comparison of the size of the set of natural numbers (1, 2, 3, 4, 5, . . .) to the set of even numbers (2, 4, 6, 8, 10, . . .). Intuitively, one would say that the set of natural numbers is much larger than the set of even numbers. However, since these are infinite sets, we can show that for every natural number there is a partner in the set of even numbers, which implies that the two sets have the same number of elements. Yet, intuitively, this is upsetting since the set of even numbers is missing all the odd numbers, which are in the set of natural numbers. Perhaps even more difficult than recognizing that the set of natural numbers is as large as the set of even numbers, is to consider that the set of real numbers (i.e., rational and irrational numbers) between 0 and 1 is larger than the set of natural numbers. The real numbers between 0 and 1 do not exist as a countable set; that is, they cannot be brought into one-to-one correspondence with the set of natural numbers. Consequently, the set of real numbers is a larger infinite set. (We will discuss uncountable sets just a bit later in this chapter.)

We have merely scratched the surface here with respect to the concept of infinity, which is typically—for obvious reasons—not presented in any substantive way at the high school level. However, we shall continue to investigate this curious concept in the following units. Although these paradoxes boggle the mind, they reveals just how fascinating the concept of infinity, and the study of mathematics, can be.

COUNTING THE UNCOUNTABLE

After having learned how to count up to 1000 and beyond, children typically start to wonder whether there exists a "largest number." They will soon recognize that there is no such thing as a largest number: Given any positive integer, we can always add 1 to produce an even larger number. Thus, the concept of a largest number is meaningless. Evidently, there exist infinitely many positive integers, so the set of natural numbers must be infinite. But there also exist other infinite sets of numbers; so let's see what happens when we try to compare two infinite sets.

Let us consider the set of all natural numbers $\mathbb{N} = \{1,2,3,\ldots\}$ and the set of all integers $\mathbb{Z} = \{\ldots,-3,-2,-1,0,1,2,3,\ldots\}$. Perhaps, we might conclude that $\mathbb{Z}$ is a substantially larger set than $\mathbb{N}$, essentially twice as large, to be more specific. This seems to be a perfectly reasonable and modest assumption that nobody would question. But since we know that $\mathbb{N}$ is infinite, our assumption of $\mathbb{Z}$ being larger than $\mathbb{N}$ implies that $\mathbb{Z}$ should be even "more infinite," in some fashion. But what should that mean? Moreover, between any two integers we can find numerous fractions. Hence the set of all fractions (or rational numbers), $\mathbb{Q}$, should be "much more infinite" than $\mathbb{Z}$, but how much more? And what about all the real numbers, $\mathbb{R}$? Is there any way to "measure" infinities?

It is surprising that those questions were addressed rather late in the history of mathematics, at least in a mathematically rigorous way. Concepts of (negative) integers, fractions, and even irrational numbers have already been used since ancient times, and philosophical discussions about infinity go back to Aristotle (384–322 BCE). Yet it took a long time until mathematical notions were invented that made it possible to compare one infinite set with another. It was the German mathematician Georg Cantor (1845–1918) who found a very simple, but brilliant, method to compare the sizes of different sets, even if they are infinite. Actually, he also defined and established the mathematical notion of a set and developed the field of set theory, which is now one of the foundations of modern mathematics.

To explain Cantor's brilliant idea for comparing sets, suppose we are given two sets, A and B, both of which contain only a finite number of elements (see figure 5.9).

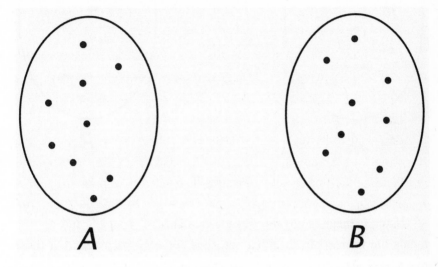

Figure 5.9

Then one (and only one) of the following three statements must be true:

1. Set A has more elements than set B.
2. Set A has fewer elements than set B.
3. Both sets A and B contain the same number of elements.

Is there any way to find out which of these statements is true without actually counting the elements of A and B? Yes, there is! We just have to pair each member of A with a corresponding member of B, for instance, by drawing a line from one to the other (see figure 5.10).

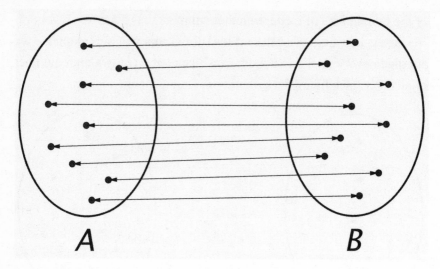

Figure 5.10

If we manage to do this for all elements of A and B and no elements are omitted from either set, then for each member of A there must be exactly one "partner" element in B, and thus both sets must contain the same number of elements. In mathematics, this is called a one-to-one correspondence between the elements of the two sets (as we discussed, briefly, earlier). Although this method of comparing sets is in fact very old, since it is actually nothing other than "counting with fingers," Cantor was the first to recognize that this strategy can also be applied to infinite sets.

The rational numbers are countable!

Cantor's notion of a one-to-one correspondence enables us to compare two infinite sets, since we don't have to actually count the number of elements in each of the sets separately, and then compare the numbers. We just need to find out whether or not we can establish a one-to-one correspondence between the elements of the two sets. Above, we tried to convince you that there are many more rational numbers (or fractions) than natural numbers. Astonishingly, this is wrong. Cantor showed that the rational numbers can be put into a one-to-one correspondence with the natural numbers. Another way of saying this is that we can give all the rational numbers a waiting number in an infinite line. We won't give a detailed proof here, but the essential idea in Cantor's proof is not too difficult to grasp. Considering only positive fractions for the moment, we can order them in a table by placing fraction $\frac{p}{q}$ in the cell at the intersection of row p and column q (see figure 5.11). For instance, the fraction $\frac{73}{111}$ will be in the table at the intersection of the 73rd row and the 111th column. Now we want to put all positive fractions in a waiting line. Of course, this waiting line will never end, since the table never ends either; but that doesn't matter. We only have to make sure that every fraction will be included. To achieve this, Cantor proposed a clever "diagonal" counting scheme (see figure 5.11): We start at $\frac{1}{1} = 1$ and draw an arrow to the right, getting to $\frac{1}{2}$. From here we move on diagonally downward to $\frac{2}{1} = 2$, then straight downward to $\frac{3}{1} = 3$, then diagonally upward, arriving at $\frac{1}{3}$ (we skipped $\frac{2}{2} = 1$ since it has already been counted). Now the whole procedure is repeated, that is, "one to the right and diagonally downward until we reach the first column, then straight down and diagonally upward." Whenever we encounter a fraction that is equivalent to one that has already gotten a number, we skip it (these are the bypassed fractions in figure 5.12).

Table of all fractions:

$\frac{1}{1}$	$\frac{1}{2}$	$\frac{1}{3}$	$\frac{1}{4}$	$\frac{1}{5}$	..
$\frac{2}{1}$	$\frac{2}{2}$	$\frac{2}{3}$	$\frac{2}{4}$	$\frac{2}{5}$	..
$\frac{3}{1}$	$\frac{3}{2}$	$\frac{3}{3}$	$\frac{3}{4}$	$\frac{3}{5}$	..
$\frac{4}{1}$	$\frac{4}{2}$	$\frac{4}{3}$	$\frac{4}{4}$	$\frac{4}{5}$	..
$\vdots$	$\vdots$	$\vdots$	$\vdots$	$\vdots$	

Figure 5.11

Table of all fractions:

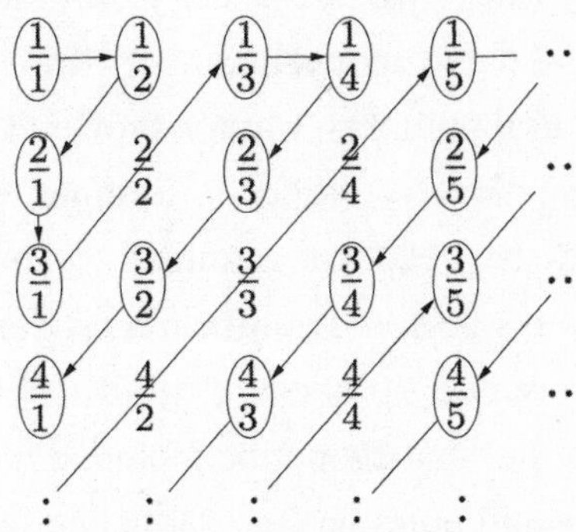

Figure 5.12

To see why a diagonal counting scheme is essential, the following picture might be helpful: Suppose you have a robotic lawnmower and the infinite table of fractions in figure 5.11 defines the area to be mowed. How should the lawnmower move in order to reach every piece of this infinite lawn? Since the infinite lawn has only one corner, it must start there and work its way in diagonal serpentines away from that corner—following the infinite waiting line drawn in figure 5.12.

By using Cantor's clever diagonal scheme, we manage to put all fractions in a waiting line without omitting any one of them. We have, therefore, established a one-to-one correspondence between all positive fractions and the natural numbers: The first fraction is $\frac{1}{1}$, the second fraction is $\frac{1}{2}$, the third one is $\frac{2}{1} = 2$, the fourth one is $\frac{3}{1} = 3$, and so on (see figure 5.12). Every fraction gets a number given by its position in the waiting line. So far we have omitted the negative fractions, but we can now simply slip each negative fraction after the corresponding positive one in the line, and place the zero at the very first position. Since we can pair each rational number with a natural number, and no numbers are omitted from either set, there must be just as many natural numbers as there are rational numbers. A set whose members can be put in a waiting line without missing any one of its elements is called a countable set. So, Cantor proved that $\mathbb{Q}$ is countable, a very surprising result!

The real numbers are uncountable!

Encouraged by this result, we might ask whether it is also possible to put the real numbers into a one-to-one correspondence with the natural numbers. Cantor showed that this is impossible, since no matter how cleverly we try to arrange the real numbers (that is, the rational and the irrational numbers) to put them in a waiting line, there will always be some numbers left over. To be precise, for any proposed list or counting procedure of all real numbers, we can always construct a number that cannot be included in this list. Real numbers can have infinite and nonrepeating sequences after the decimal point. It is this property that makes them "uncountable." An uncountable set is a set that contains too many elements to be countable and is therefore "larger" than the set of natural numbers, $\mathbb{N}$. Instead of showing that the set $\mathbb{R}$ is uncountable, it suffices to show that an arbitrary subset of $\mathbb{R}$ is already uncountable. This means, if we can show for any concrete selection of real numbers that they are too many to be countable, then the whole set of real numbers, $\mathbb{R}$, must be uncountable as well. Following Cantor's 1891 proof, we shall consider only real numbers between 0 and 1. In addition, we take only those real numbers between 0 and 1 into account, whose fractional part is non-terminating and consists exclusively of zeros and ones. Notice that the largest number satisfying these requirements is $0.\overline{1}$. Now suppose somebody claims to have found a procedure to enumerate all such numbers, and then presents a list of all fractional parts that are possible under these constraints. Such a list might look like the list shown in figure 5.13 (which, for space purposes, provides only the first six numbers of the list).

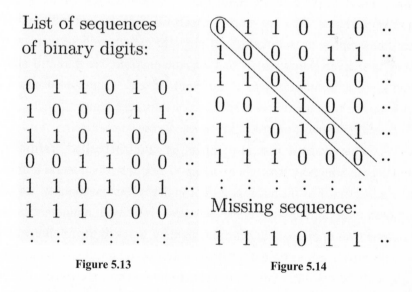

List of sequences of binary digits:						
0	1	1	0	1	0	..
1	0	0	0	1	1	..
1	1	0	1	0	0	..
0	0	1	1	0	0	..
1	1	0	1	0	1	..
1	1	1	0	0	0	..
:	:	:	:	:	:	

Figure 5.13

0	1	1	0	1	0	..
1	0	0	0	1	1	..
1	1	0	1	0	0	..
0	0	1	1	0	0	..
1	1	0	1	0	1	..
1	1	1	0	0	0	..
:	:	:	:	:	:	

Missing sequence:

1 1 1 0 1 1 ..

Figure 5.14

We show that we can always write down a sequence of zeros and ones that is not a member of this list. We take the first digit of the first sequence in the list and write down its complement (that means change 0 to 1 and 1 to 0). For the next digit, take the complement of second digit of the second sequence, then the complement of the third digit of the third sequence, and so on (see figure 5.14). By construction, this "diagonal sequence" differs from all sequences in the list, because it was constructed in a manner that its digit at position n does not coincide with the corresponding digit of the nth number in the list. It must be different from the first sequence in the list, because its first digit is different. It must also be different from the second sequence in the list, because its second digit is different. It must be different from the third sequence in the list, because its third digit is different, and so on. Therefore, this sequence cannot occur in the enumeration! This proof is now known as Cantor's diagonal argument, and the construction of a "diagonal sequence" from an infinite set of sequences became an important technique that is frequently used in mathematical proofs.

Cantor showed that it is impossible to enumerate the real numbers; they cannot be put into a one-to-one correspondence with the natural numbers. Hence, they are more "numerous" than the natural numbers and thus represent a "greater" infinity. He called such sets uncountable sets. Even though such sets are uncountable, we can still show that there exist infinite sets of different "sizes." Cantor also showed that among uncountable sets there exist infinitely many of them, and he developed an arithmetic of infinities. To measure the size of infinite sets, he extended the natural numbers by numbers called "cardinals" and denoted them by the Hebrew letter $\aleph$ (aleph) with a natural number as a subscript. For instance, $\aleph_0$ (aleph-null) is the "cardinality" of the set of natural numbers—it is the "smallest" infinity in mathematics. At the time Cantor published these results, they shocked the mathematical community. They were in conflict with common beliefs and were considered to be revolutionary. Many renowned mathematicians even tried to prove Cantor wrong and did not accept his work. The criticism of his work threw him into a depression, and he even gave up mathematics for some time. Although he recovered and continued his scientific work, he never fully regained his passion for mathematics. It took decades until the importance and ingenuity of his ideas was fully recognized. Cantor was ahead of his time.

Thus, we have shown that the set of rational numbers, $\mathbb{Q}$, is not larger than the set of natural numbers, $\mathbb{N}$, which is a totally counterintuitive fact. Although this statement seems to contradict common sense, its proof is actually rather simple and not very difficult to follow. The same is true for the proof of the set of real numbers, $\mathbb{R}$, being essentially larger than $\mathbb{N}$, showing that there exist mathematical infinites of different sizes. That very surprising and unexpected results can be found among the most innocent structures such as the natural, rational, and real numbers is one facet of the beauty of mathematics.

MATHEMATICS ON A BICYCLE

For a change of pace, let us now consider an application from everyday life that could be presented in the school curriculum but rarely ever is. With the increased proliferation of bicycles in our society today—especially in the big cities—it would be good to see how mathematics can help us better understand gear selection. Many bicycles have multiple gears and thus multiple gear ratios. A shifting mechanism allows us to select an appropriate gear ratio for efficiency or comfort under the prevailing circumstances. The bicycles that we shall consider have two wheels of equal diameters and derailleur gears with one, two, or three sprocket wheels in the front and a sprocket wheel cluster on the rear wheel. The latter typically consists of five or more sprocket wheels, depending on the type of the bicycle. On the rear wheel, the largest sprocket wheel is closest to the spokes and then the rest of the sprocket wheels are arranged such that the smallest is outermost. (See figure 5.15.)[3]

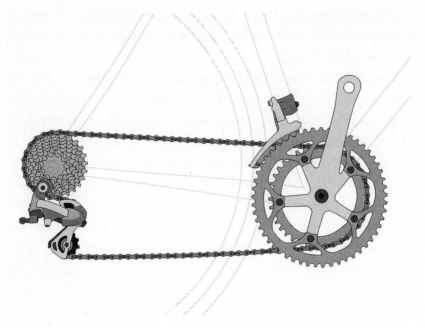

Figure 5.15

The sprocket wheels in the front are also called chainrings. They are attached to the cranks to which the pedals attach. The gearing (that is, the connection of sprocket wheels by a chain) is obtained by moving the chain from one sprocket to the other by means of a derailleur, a device that lifts the driving chain from one sprocket wheel to another of different size.

Modern racing bikes often have 10-speed or 11-speed cassettes (sprocket clusters) and two chainrings in the front. Thus, there are up to twenty-two different gears possible. Mountain bikes, on the other hand, usually have three chainrings in the front and, accordingly, up to thirty-three different gears. This is just a theoretical maximum, since some gears would result in a very diagonal chain alignment and should not be used, as they would cause excessive chain wear. Bicycles for off-road cycling need very low gears for steep grades with poor traction. That's why mountain bikes have a third, particularly small, chainring in the front.

Let's now examine closely the basic setup. There exists a front and rear sprocket wheel with teeth set in gear by a connecting chain. The numbers of teeth on the front and rear sprocket wheels are important. Suppose the front sprocket wheel has forty teeth and the rear sprocket wheel has twenty teeth; the ratio would then be $\frac{40}{20}$, or 2. This means that the rear sprocket wheel turns twice every time the front sprocket wheel turns once. But the rear sprocket wheel is attached to the bicycle wheel and so the bicycle wheels will turn twice as well. The distance traveled during one complete revolution of the pedals also depends on the diameter of the drive wheel. Thus, including the bicycle wheel in our consideration, the relevant quantity is the gear ratio measured in inches, also known as gear inches. The gear ratio in inches is defined as the product of the diameter of the drive wheel in inches and the ratio between the number of teeth in the front chainring and the number of teeth in rear sprocket. The result is usually rounded to the nearest whole number.

$$gear\ inches = bicycle\ wheel\ diameter \cdot \frac{number\ of\ teeth\ in\ the\ front\ chainring}{number\ of\ teeth\ in\ the\ rear\ sprocket}$$

Assuming that the diameter of the wheel (including the tire) is 27 inches, we would have $2 \cdot 27'' = 54$ gear inches for a 40-tooth chainring and a 20-tooth sprocket. The number generated here can range from 20 (very low gear) to 125 (very high gear). It gives a comparison between gears, and bicycle experts and racers use it to fine-tune the gearing to better suit their level of fitness, their muscle strength, and the expected use. If you multiply the gear inches obtained in the formula above by π, you get the distance traveled forward with each turn of the pedals, which is called "meters of development." (Recall: circumference $= \pi \cdot$ diameter.) A higher gear is harder to pedal than a lower gear, since the distance traveled during one revolution of the pedals is longer and therefore the work performed by the rider with each turn of the pedals must be larger.

For example, a rider using a sprocket wheel with 46 teeth in the front and a 16-tooth wheel in the rear along with a 27" wheel gets a gear ratio of 77.625 ≈ 78. Another rider using a 50-tooth front-sprocket wheel and 16-tooth sprocket wheel in the rear gets a gear ratio of 84.375 ≈ 84, which would be harder to pedal than a 78-gear ratio. The rider with the 78-gear ratio goes approximately 245 inches (or 6.2 meters) forward for each turn of the pedals; the rider with the 84-gear ratio goes approximately 264 inches (or 6.7 meters) forward for each complete turn of the pedals. Hence, the increased difficulty in pedaling is returned in greater distance per pedal revolution.

Now let us examine applications of various gearing ratios to the average rider, whom we will call Charlie. Suppose Charlie was riding comfortably on level ground in a 78-gear ratio, and then he came to a rather steep hill. Should he switch to a higher or lower gear ratio? Your reasoning should be as follows: If Charlie switches to an 84-gear ratio, he will go 264 inches forward for each turn of the pedals. This requires a certain amount of work. To overcome the effects of gravity to get up the hill requires additional energy. So Charlie would probably end up walking his bicycle. If Charlie had switched to a lower gear, he would use less energy to turn the pedals and the additional energy required to climb the hill would make his gearing feel about the same as the 78-gear ratio. So the answer is to switch to a lower number gear ratio. Charlie will have to turn the pedal more revolutions to scale the hill; more than if he had chosen the 84-gear ratio, and more than if he stayed in the 78-gear ratio. Remember, his gearing only *feels* like 78 because of the additional work required to climb the hill.

The work performed while riding a bicycle from A to B is independent of the chosen gear, but a rider can adapt the gear to the slope of the road and to her or his physical condition. The trade-off for reducing the pedaling force by switching to a lower gear is a higher number of pedal revolutions that are necessary to get from point A to point B. However, the force exerted on the pedals and the total number of revolutions are not the only relevant criteria. A cyclist's legs produce power optimally within a narrow pedaling-speed range, or cadence. The cadence is the number of pedal revolutions per minute (rpm). For a bicycle to travel at the same speed (or in the same time from A to B), using a lower gear requires the rider to pedal at a faster cadence but with less force. Conversely, a higher gear provides a higher speed for a given cadence but requires the rider to exert greater force. For professional cyclists, choosing the right gear is essentially a matter of finding the optimal balance between cadence and pedaling force. In a bicycle race, for instance, a stage of the Tour de France, the cyclists will most of the time choose the gear such that the cadence always remains in their preferred corridor. The preferred cadence varies individually and is somewhere between 70 and 110 rpm for most cyclists. The following table shows the bicycle's speed for different gears and cadences.

Gear	Gear Inches	Meter Development	Front Teeth/ Rear Teeth	60 rpm mph	60 rpm km/h	80 rpm mph	80 rpm km/h	100 rpm mph	100 rpm km/h
Very high	125	10	53/11	22.3	36	29.7	47.8	47.1	59.7
High	100	8	53/14	18	29	24	38.6	30	48.3
Medium	70	5.6	53/19 or 39/14	12.5	20	16.6	26.7	21	33.6
Low	40	3.2	34/23	7.2	11.6	9.6	15.4	11.9	19.2
Very low	20	1.6	32/42	3.5	5.6	4.7	7.6	5.9	9.5

Of course, there are circumstances in which we must leave the preferred cadence. This is typically the case in the final sprint of a race. To reach maximum speed, a cyclist will always select the highest gear and pedal as fast as possible. On the other hand, when riding up a very steep hill (for example, the notorious Alpe d'Huez climb used in the Tour de France), the cadence might fall substantially below 70 rpm if the cyclist already uses the lowest gear, but the slope of the road is too high to maintain the preferred cadence.

A racer will carefully select his or her back sprocket-wheel cluster depending on the course. A relatively flat course would necessitate a tooth range of 11–23 in the rear sprocket-wheel cluster, while a mountain course would require lower gears to climb up the hill and thus, for instance, a tooth range of 11 – 32 (the 11-tooth sprocket is kept for potential sprint finishes). Finally, preventing possible gear ratio duplication will also affect the choice. In the table, we saw duplication of the same gear ratio with a 14- and a 19-tooth rear sprocket wheel (and chainrings with 53 and 39 teeth). This reduces the number of different gears and should be avoided. Duplicated or near-duplicated gear ratios occur on many less-expensive bicycles. Another aspect to be taken into consideration is the increments in gear inches from one gear to the next. The relative change from a lower gear to a higher gear can be expressed as a percentage. Cycling tends to feel more comfortable if nearly all gear changes have more or less the same percentage difference. For example, a change from a 13-tooth sprocket to a 15-tooth sprocket (15.4 percent) feels very similar to a change from a 20-tooth sprocket to a 23-tooth sprocket (15 percent), even though the latter has a larger absolute difference. Consistent percentage differences can be achieved if the number of teeth are in geometric progression. Now that we have seen how the circle dominates the operation of a bicycle, we will transition to another very popular curve, which *is* taught in high school curriculum, but unfortunately lacks some motivational applications. These we will now investigate.

THE PARABOLA: A REMARKABLE CURVE

With a ruler and a compass, we can draw two sorts of "lines"—straight lines and circles, or circular arcs. We put the word "line" in quotation marks because in mathematics the notion of a line is synonymous to "straight line." "Curve" is the more general term for a "line" that is not necessarily straight. Straight lines and circular arcs are special examples of curves. They are the most basic curves and often the only curves that are drawn in elementary plane geometry. It is obvious that studying the geometrical properties of shapes constructed from these basic types of curves represents useful knowledge. Basic shapes constructed from straight lines and circular arcs include triangles, quadrilaterals, polygons, complete circles, semicircles, circular segments, and so on—which can be drawn by using only a ruler and a compass. However, there exist many other curves as well. Although we cannot draw them by using a ruler and a compass alone, they are no less important for applications in technology. The parabola is an example of a curve on whose properties we rely every time we watch satellite television. Receiving electromagnetic signals from satellites requires a "satellite dish," which is a special antenna designed to receive electromagnetic signals from satellites. A satellite dish is essentially a dish-shaped parabolic antenna with a device that amplifies the received signals and converts them into electrical currents, which are then transmitted to the satellite receiver. The very shape of the satellite dish is crucial, but before exposing the reason for this property, we ought to briefly recall the definition of a parabola.

A parabola is defined by a fixed straight line, d (the directrix), and a fixed point, F, not on this line (the focus). The parabola is the set of all points in the plane that are equidistant from d and F. This means that a point belongs to the parabola if and only if its distance to the focus, F, is equal to its distance to the directrix, d. (Recall that the distance of a point to a line is the length of the perpendicular segment from the line to the point.) One such point is easily obtained by drawing a line perpendicular to d through F and bisecting the segment from d to F (see figure 5.16).

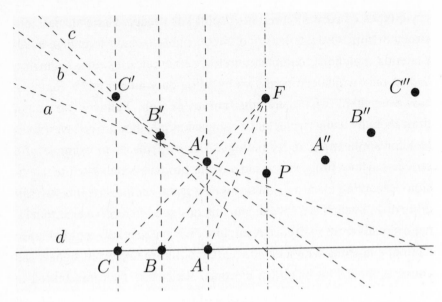

Figure 5.16
The construction of a parabola.

More points of the parabola can be constructed as follows. Take any
point, A, on the directrix, d, and draw the perpendicular bisector of the
segment AF, which we denote by a. Then all points on a are equidistant to
A and F. Finally, we construct a line through A perpendicular to d, intersecting a at point A'. Then A' will be a point of the parabola, since it has
equal distance to F and d. By construction, the distance between A' and d
is precisely equal to the distance between A' and A. Moreover, since A' lies
on a, it is equidistant to F and A and therefore to F and d. To get points of
the parabola on the other side of FP, we could continue this process, or
we could reflect point A' about the axis of symmetry FP, whose reflected
point is denoted by A''. Now we may pick other points, B and C, on d and
construct B' and B'' as well as C' and C'', respectively (see figure 5.16).
Repeating this procedure for all (or many) points on d, we would get all
(or many) points of the parabola. Note that the bisectors of line segments
can be created in paper by folding one end of the segment onto the other.
For this reason, tangents of parabolas are created by folding the points of

the directrix onto the focus. This is the reason why parabolas are the basic curves in the intrinsic geometry of paper folding, origami. Much more on this fascinating topic can be discovered in the book *Geometric Origami* by Robert Geretschläger (Shipley, UK: Arbelos Publishing, 2008).

Let's go back to our example of the satellite dish. The three-dimensional shape of a satellite dish emerges if we rotate the parabola (see figure 5.17) around its axis of symmetry (see figure 5.18). The technical term for the surface generated by rotating a parabola around its symmetry axis is *paraboloid*. The reason for using this shape for satellite dishes is a wonderful geometric property of paraboloids. Imagine a mirror in the shape of a paraboloid (the reflecting layer being inside the paraboloid) and light rays hitting the mirror parallel to the paraboloid's axis of symmetry (that is, the perpendicular line from the focal point to the directrix). Then all the light would be reflected into the focal point, where the intensity will, thus, be very high, meaning that a parabolic mirror acts as an amplifier for light. But light is actually an electromagnetic wave, and so are television signals from satellites. Light rays are just a way to represent the direction in which the wave is traveling. A parabolic antenna amplifies the incoming signals by reflecting them into the focal point where a device is mounted that converts the signal to an electrical current.

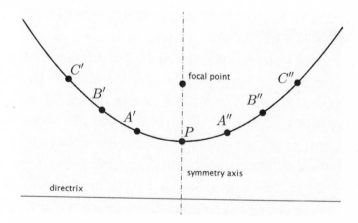

Figure 5.17
The parabola defined by its directrix and its focal point.

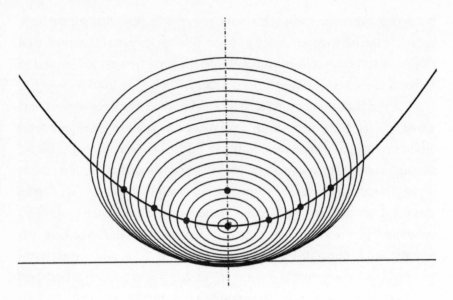

Figure 5.18
A paraboloid is obtained by rotating the parabola
around its symmetry axis.

Before exploring the reflective property of the parabola, we have to recall the law of reflection, which states that the angle of incidence is equal to the angle of reflection measured from the normal. The normal is perpendicular to the surface, that is, perpendicular to the plane tangent to the surface (see figure 5.18).

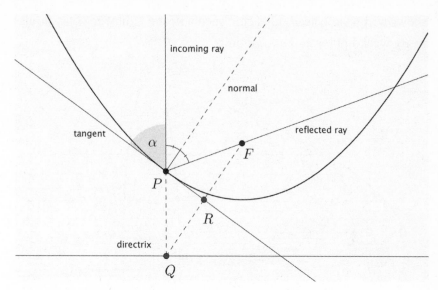

Figure 5.19
Light rays coming in parallel to the symmetry axis
of the parabola converge at the focal point.

Now let us consider a ray hitting the parabola at point P at an angle α with the tangent to the parabola, as shown in figure 5.19. We extend the ray to the directrix and denote the point of intersection by Q. Moreover, we draw the tangent to the parabola through P. Then $PQ = PF$ by the definition of a parabola, and PR is the altitude of the isosceles triangle QPF. The angle FPR is therefore congruent to angle QPR. But $\angle QPR = \alpha$, since vertical angles are congruent. Hence, PF indeed represents the direction of the ray reflected from the parabolic mirror. Since P was an arbitrary point on the parabola, this must be true for all points on the parabola, meaning that a beam of electromagnetic waves coming in parallel to the symmetry axis will converge at the focal point, F.

Finally, since the law of reflection is symmetrical, the opposite is also true. Light emitted by a source placed at the focal point will be focused outward along the direction of the symmetry axis. This is exactly how flashlights or car headlights work. See figure 5.20, which

shows why a car's headlights emit such a strong light when only a relatively weak light is at the focal point.

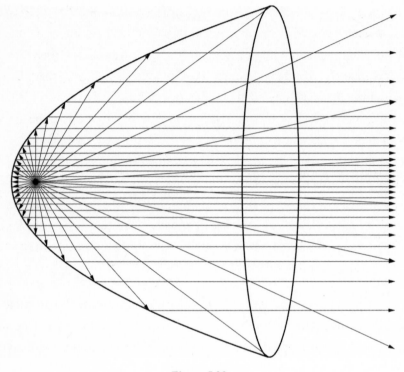

Figure 5.20

As shown in figure 5.20, when light rays emitted by a source at
the focal point are reflected inside the parabolic headlight, they will all
leave the headlight parallel to its axis of symmetry, resulting in a strong
concentration of light in this direction.

We have now come to the end of the book with the hope that you have
had a chance to experience mathematics as a subject that not only is
exciting in and of itself, but also has many applications of the material that is taught in school and unfortunately too often omitted. Many
of these "omissions" can help explain lots of the things we take for

ACKNOWLEDGMENTS

The authors wish to acknowledge superb support services received from the publisher Prometheus Books, led by their editor in chief, Steven L. Mitchell, and his truly dedicated production coordinator, Catherine Roberts-Abel. We wish to also thank Senior Editor Jade Zora Scibilia for her highly meticulous editing and clever suggestions to make the presentation as intelligible as possible. Thanks is also due to Editorial Assistant Hanna Etu, and the typesetter, Bruce Carle. The cover design exhibits the talents of Nicole Sommer-Lecht. We're also very pleased with the indexing by Laura Shelley.

Each of the authors has many people to thank for their patience and support throughout this book-development process. In particular, Dr. Christian Spreitzer wants to thank Katharina Brazda for inspiring discussions, which resulted in some very creative contributions.

APPENDIX

As we mentioned earlier, Ceva's theorem might well have been introduced to a high school class, since it merely applies similarity relationships that are an integral part of the geometry curriculum. We offer one of many proofs available to justify Ceva's theorem. It is perhaps easier to follow the proof by looking at the left-side diagram in figure App.1 and then verifying the validity of each of the statements in the right-side diagram. In any case, the statements made in the proof hold for *both* diagrams.

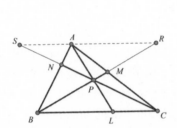 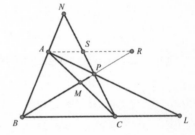

Figure App.1.

Consider figure App.1, for which we have on the left triangle *ABC* with a line (*SR*) containing *A* and parallel to *BC*, intersecting *CP* extended at *S* and *BP* extended at *R*.

The parallel lines enable us to establish the following pairs of similar triangles:

$$AMR \sim CMB; \text{ therefore, } \frac{AM}{MC} = \frac{AR}{CB}. \tag{I}$$

$BNC \sim ANS$; therefore, $\dfrac{BN}{NA} = \dfrac{CB}{SA}$.   (II)

$CLP \sim SAP$; therefore, $\dfrac{CL}{SA} = \dfrac{LP}{AP}$.   (III)

$BLP \sim RAP$; therefore, $\dfrac{BL}{RA} = \dfrac{LP}{AP}$.   (IV)

From (III) and (IV) we get $\dfrac{CL}{SA} = \dfrac{BL}{RA}$.

This can be rewritten as $\dfrac{CL}{BL} = \dfrac{SA}{RA}$.   (V)

Now by multiplying (I), (II), and (V), we obtain our desired result:

$$\frac{AM}{MC} \cdot \frac{BN}{NA} \cdot \frac{CL}{BL} = \frac{AR}{CB} \cdot \frac{CB}{SA} \cdot \frac{SA}{RA} = 1.$$

This can also be written as $AM \cdot BN \cdot\cdot CL = MC \cdot NA \cdot BL$. A nice way to read this theorem is that the product of the alternate segments along the sides of the triangle made by the concurrent line segments (called cevians) emanating from the triangle's vertices and ending at the opposite side are equal.

Yet, it is the converse of this proof that is of particular value to use here. That is, if the products of the alternate segments along the sides of the triangle are equal, then the cevians determining these points must be concurrent.

We shall now prove that if the lines containing the vertices of triangle ABC intersect the opposite sides in points L, M, and N, respectively, so that $\frac{AM}{MC} \cdot \frac{BN}{NA} \cdot \frac{CL}{BL} = 1$, then these lines, AL, BM, and CN, are concurrent.

Suppose BM and AL intersect at P. Draw PC and call its intersection with AB point N'. Now that AL, BM, and CN' are concurrent, we can use the part of Ceva's theorem proved earlier to state the following:

$$\frac{AM}{MC} \cdot \frac{BN'}{N'A} \cdot \frac{CL}{BL} = 1.$$

But our hypothesis stated that $\frac{AM}{MC} \cdot \frac{BN}{NA} \cdot \frac{CL}{BL} = 1$.

Therefore, $\frac{BN'}{N'A} = \frac{BN}{NA}$, so that N and N' must coincide, which thereby proves the concurrency.

For convenience, we can restate this relationship as follows: If $AM \cdot BN \cdot CL = MC \cdot NA \cdot BL$, then the three lines are concurrent.

The ambitious reader may want to see other proofs of Ceva's theorem, which can be found in *Advanced Euclidean Geometry* by Alfred S. Posamentier (New York: John Wiley and Sons, 2002, pp. 27–31.

NOTES

INTRODUCTION

1. Michel Chasles, *Aperçu historique* 2 (1875).

CHAPTER 1: ARITHMETIC NOVELTIES

1. See MacTutor History of Mathematics Archive, School of Mathematics and Statistics, University of St Andrews, Scotland, "An Overview of Babylonian Mathematics," http://www-history.mcs.st-and.ac.uk/HistTopics/Babylonian_mathematics .html and the references therein.

2. Pi World Ranking List, http://www.pi-world-ranking-list.com/index.php ?page=lists&category=pi.

3. Proper divisors are all the divisors, or factors, of the number except the number itself. For example, the proper divisors of 6 are 1, 2, and 3, but not 6.

4. If $k = pq$, then $2^k - 1 = 2^{pq} - 1 = (2^p - 1)(2^{p(q-1)} + 2^{p(q-2)} + \cdots + 1)$. Therefore, $2^k - 1$ can only be prime when k is prime, but this does not guarantee that when k is prime, $2^k - 1$ *will* also be prime, as can be seen from the following values of k:

k	2	3	5	7	11	13
$2^k - 1$	3	7	31	127	2,047	8,191

Note that $2,047 = 23 \cdot 89$ is not a prime and so doesn't qualify.

5. Rounded to nine decimal places.

CHAPTER 2: ALGEBRAIC EXPLANATIONS
OF ACCEPTED CONCEPTS

1. Kurt von Fritz, "The Discovery of Incommensurability by Hippasus of Metapontum," *Annals of Mathematics* 46, no. 2, 2nd ser. (April 1945); see also chap. 6 in *Lore and Science in Ancient Pythagoreanism*, by Walter Burkert (Cambridge, MA: Harvard University Press, 1972).

CHAPTER 3: GEOMETRIC CURIOSITIES

1 . L. Hoehn, "A Neglected Pythagorean-Like Formula," *Mathematical Gazette* 84 (March 2000): 71–73.

2. Deanna Haunsperger and Stephen Kennedy, eds., *The Edge of the Universe: Celebrating 10 Years of Math Horizons* (Washington, DC: Mathematical Association of America, 2006), p. 231.

3. This reference to *pons asinorum* would appear to be wrong, since we usually consider the proof that the base angles of an isosceles triangle are congruent as the pons asinorum, or "bridge of fools." It is clear that they meant to refer to the Pythagorean theorem. *New England Journal of Education* 3, no. 14 (April 1, 1876).

4. Elisha S. Loomis, *The Pythagorean Proposition* (Reston, VA: NCTM, 1940, 1968).

5. A book to consider is *The Pythagorean Theorem: The Story of Its Power and Beauty*, by A. S. Posamentier (Amherst, NY: Prometheus Books, 2010).

6. This idea was published by Cabre Moran, "Mathematics without Words," *College Mathematics Journal* 34 (2003): 172.

7. For more information about Ceva's theorem and related topics, see *The Secrets of Triangles*, by A. S. Posamentier and I. Lehmann (Amherst, NY: Prometheus Books, 2012).

8. "Small stellated dodecahedron" was created by Robert Webb's Stella Software, http://www.software3d.com/Stella.php.

9. "Great stellated dodecahedron" was created by Robert Webb's Stella Software, http://www.software3d.com/Stella.php.

10. "Great dodecahedron" was created by Robert Webb's Stella Software, http://www.software3d.com/Stella.php.

11. "Great icosahedron" was created by Robert Webb's Stella Software, http://www.software3d.com/Stella.php.

CHAPTER 4: PROBABILITY APPLIED TO EVERYDAY EXPERIENCES

1. Interested readers might want to read the book by Jason Rosenhouse, *The Monte Hall Problem: The Remarkable Story of Math's Most Contemptuous Brainteaser* (New York: Oxford University Press, 2009).

2. For readers interested in methods of calculating the number of possible hands of a certain type, here is a brief introduction to the procedures you can use to get these numbers.

Here is how to calculate the number of three-of-a-kind hands when there are two jokers in the deck:

Out of the

$$\begin{pmatrix} 54 \\ 5 \end{pmatrix}$$

possible hands that can be dealt from a deck of 52 cards plus two jokers, there are three types of hands that will normally be counted as three of a kind. These can include 0, 1, or 2 jokers.

First of all, let us consider how many three-of-a-kind hands there are that do not include a joker. The three cards of the same kind can have any of thirteen values, from a 2 through an ace. There are four ways to choose three of these (as one of the four will not be chosen). Furthermore, the other two cards can be of any of the remaining twelve values, and there are

$$\begin{pmatrix} 12 \\ 2 \end{pmatrix} = 66$$

such choices possible. Each of these values can be represented by any of the four suits. In total, there are, therefore,

$$13 \cdot 4 \cdot \begin{pmatrix} 12 \\ 2 \end{pmatrix} \cdot 4^2 = 54{,}912$$

such hands.

Next, let us consider how many three-of-a-kind hands include exactly one joker. These hands all result from one pair, two other different cards, and the joker to complete the three of a kind together with the pair. We obtain numbers in a similar fashion to the previous case. The pair can have any of thirteen values, from a 2 through an ace, and there are

$$\left(\begin{array}{c} 4 \\ 2 \end{array} \right) = 6$$

ways to choose two of these from the four cards of this value. Furthermore, the other two cards can once again be of any of the remaining twelve values, and there are again

$$\left(\begin{array}{c} 12 \\ 2 \end{array} \right) = 66$$

such choices possible, with each of these values being represented by any of the four suits. Finally, there are also 2 possible jokers in play, and either of these can be part of the hand. In total, there are therefore,

$$13 \cdot \left(\begin{array}{c} 4 \\ 2 \end{array} \right) \cdot \left(\begin{array}{c} 12 \\ 2 \end{array} \right) \cdot 4^2 \cdot 2 = 164,736$$

such hands. (Note that this is three times as many as there are "normal" three of a kind hands.)

Finally, we have the most difficult case to consider, the case in which two of the cards in the hand are jokers. Since we have only three more cards to choose, this may seem to be easier, but this is not the case. First, we note that the three cards must be of different values, since any pair would make four of a kind possible, which would yield a more valuable hand than three of a kind. These three may not be of the same suit, since we could otherwise define the two jokers to complete a flush, which is also more valuable than three of a kind, and they may also not be within the same sequence of five values (since we could otherwise define the two jokers to complete a straight). Also, in eliminating these two possibilities, we must note that a straight flush (or royal flush) is also an option, and this is included in both, so simple subtraction of these cases will not suffice. Since these hands will have been subtracted twice, we can add their number again once (an application of the so-called *inclusion-exclusion principle*) to obtain the number we want. This gives us the calculation:

$$\left(\begin{array}{c} 13 \\ 3 \end{array} \right) \cdot 4^3 - 4 \cdot \left(\begin{array}{c} 13 \\ 3 \end{array} \right) - 64 \cdot 4^3 + 64 \cdot 4 = 13320,$$

and therefore, the total number of three-of-a-kind hands is 54912 + 164736 + 13320 = 232968. (Note that the number 64 for the number of possible combinations leading to a straight results from allowing ace high or low, which is not always applied.)

3. See S. Gadbois, "Poker with Wild Cards—A Paradox?" *Mathematics Magazine* 69 (1996): 283–85.

CHAPTER 5: COMMON SENSE FROM A MATHEMATICAL PERSPECTIVE

1. Florian Cajori, *A History of Mathematical Notations*, vol. 2 (Chicago: Open Court Publishing, 1929), pp. 182–83.

2. This triangle is formed by beginning on top with 1, then the second row has 1, 1, then the third row is obtained by placing 1s at the end and adding the two numbers in the second row (1 + 1 = 2) to get the 2. The fourth row is obtained the same way. After the end 1s are placed, the 3s are gotten from the sum of the two numbers above (to the right and left), that is, 1 + 2 = 3, and 2 + 1 = 3.

3. Image from Wikimedia Creative Commons; author: Keithonearth; licensed under CC BY-SA 3.0 via Commons, https://commons.wikimedia.org/wiki/File :Derailleur_Bicycle_Drivetrain.svg#/media/File:Derailleur_Bicycle_Drivetrain.svg.

INDEX

NUMBERS

SYMBOLS AND BASIC EQUATIONS